Here's what recruits are saying abou

The information in this book is not only well researched, but also very helpful for any firefighter recruit. I can personally vouch for the interviewing section in the book as it has helped me to get recruited by the Sarnia Fire Department. This book is definitely worth the money.

~ *Shawn Schinkel*

I got recruited from Mississauga Fire Department yesterday afternoon. "I'm on cloud nine!" I wanted to drop you a line and say, "thanks for all your help." You and your website have been a big help in achieving my dream. Thanks again!

~ *Dan Burton*

It's so great to have another resource to double check things besides word of mouth. While trying to update my résumé, I really appreciated all the help from the résumé review, templates, and evaluator. All these tools will benefit me in the recruiting process.

~ *Jessica Thomson*

This guide will help you organize everything you need to apply for a firefighter position. Before I got this book, I had no clue about what to put in my application package and what order it should go in, and my résumé was horrible. However, the book and the website www.becomingafirefighter.com changed all that for me by reviewing my résumé, cover letter, and reference letter, making it 10 times better than before I got the membership … This isn't even the half of it. Think about the amount of money you have spent or will spend on your firefighting training, application fees, fitness tests, and aptitude tests. Do yourself a favor and invest in this book. It's worth it.

~ *Scott Tadema*

THE COMPLETE GUIDE TO
BECOMING
A FIREFIGHTER

REVISED AND UPDATED

THE COMPLETE GUIDE TO
BECOMING
A FIREFIGHTER

REVISED AND UPDATED

Kory Pearn

From the Front Line of Firefighting

Fitzhenry & Whiteside

Published in Canada by Fitzhenry & Whiteside, 195 Allstate Parkway, Markham, Ontario L3R 4T8

Published in the United States by Fitzhenry & Whiteside,
311 Washington Street, Brighton, Massachusetts 02135

www.fitzhenry.ca godwit@fitzhenry.ca
10 9 8 7 6 5 4 3 2 1

Library and Archives Canada Cataloguing in Publication
Pearn, Kory
The complete guide to becoming a firefighter / Kory Pearn.
— Rev. and updated ed.
ISBN 978-1-55455-436-2
1. Fire fighters—Vocational guidance—Canada. I. Title.
TH9119.P43 2010   363.37'802371   C2010-905749-X

U.S. Publisher Cataloging-in-Publication Data
(Library of Congress Standards)
Pearn, Kory.
The complete guide to becoming a firefighter / Kory Pearn.
Rev and updated.
Originally published: Toronto: White Knight Books, 2006.
[220] p. : ill. ; cm.
ISBN-13: 978-1-55455-436-5 (pbk.)
1. Fire extinction — Vocational guidance. I. Title.
363.37023 dc22   TH9119.P43 2010

Fitzhenry & Whiteside acknowledges with thanks the Canada Council for the Arts, and the Ontario Arts Council for their support of our publishing program. We acknowledge the financial support of the Government of Canada through the Book Publishing Industry Development Program (BPIDP) for our publishing activities.

Cover and interior design by Intuitive Design International Ltd.
Cover image courtesy ©2010 Fotosearch
Printed in Canada

DEDICATION

This book is dedicated to my two beautiful girls, Isabella and Eliana Pearn. May you always have the courage to follow your dreams. I love you both with all my heart. Love, Dad.

ACKNOWLEDGMENTS

This book would not have been possible without the love and support that my wife, Lisa, has unflaggingly given throughout this project. Not only is she my cheering section, but also my inspiration.

To all my family—who have always been there. Thanks.

Thank you to the St. Thomas Fire Department for allowing me the freedom to involve the fire department as well as members of the fire department in this project. As to those members—Ian Thomas, Scott Brett, Chris Gielen, and Daryl Smith—thanks for your contributions and encouragement.

Special thanks to Jeanette Fletcher and Michael Stoparczyk from Studio 19 Photography and Digital Imaging for capturing the images that appear throughout the book.

CONTENTS

ABOUT THE BOOK

Becoming a firefighter is a dream career for many men and women. This book is a straightforward guide for turning your dream into reality. I spent three years writing this book after being recruited by the St. Thomas Fire Department. Not only have I researched this material—I have lived it!

This book does not retrace the steps I took to become a firefighter, but instead acts as a resource tool that guides the reader through the process. Every step to becoming a firefighter is included, allowing the reader to start reading at any stage in their journey. The book also features many practical exercises, which not only provide information to the reader, but also enable the reader to start to use the information right away.

HOW TO GET THE MOST
OUT OF THIS BOOK

This book is a resource tool full of ideas, examples, tips, and suggestions to inspire you and to provide vital information that will allow you to be in control of your own recruitment process. Here's how you can get the most out of the material in this book:

- Adopt the ideas but make them your own. Every firefighter recruit is unique in the sense that everyone has different qualities to offer fire departments. Choose the ideas that will work for you and adapt them to suit your profile.

- You will find many new ideas in this book—don't get overwhelmed and try to accomplish everything at once. Maintain your focus once you have figured out a plan, and stick to it. Planning is the first and most critical step in any recruitment process.

- Reading the book once from cover to cover will not be the end of it. Refer back to various sections of the book before each stage of your recruitment process. Refresh your memory with the tips and ideas you'll find in each chapter.

- Most important, be sure to continually evaluate your recruitment process. Ask yourself if there is something more you could be doing. Are you on the right path? What are other recruits doing? Have you been making progress?

- Make notes right on the pages and / or highlight the passages that interest you. Carry this book around so you can refer to it at any time.

For additional information on cover letters, résumés, recruitment trends, and recruiting fire departments, www.becomingafirefighter.com is an excellent interactive recruitment website. Both the book and the website will become invaluable tools if you are a serious firefighter recruit. Kory Pearn, the author of *The Complete Guide to Becoming a Firefighter*, manages this website. Members of the website receive vital up-to-date information and are welcome to contact him at any time with questions or concerns regarding their recruitment.

Become a member of:

www.becomingafirefighter.com

Part I

Getting Started

CHAPTER ONE

JOB DESCRIPTION OF TODAY'S FIREFIGHTER

One of the most common phrases used to describe firefighters is *"They're always running into a burning building when others are running out!"* You see this slogan on T-shirts and in the media.

The constant physical demands of the job are great, and firefighters must call upon years of education and training. Today's firefighter is trained to handle every type of emergency situation, including motor vehicle accidents, train derailments, hazardous material incidents, home accidents, and of course fires, though only a small percentage of the emergency calls firefighters respond to are in fact fires. The rest are medical-related emergencies: heart attacks, strokes, VSAs (vital signs absent), seizures, severe bleeding, drowning, and any other emergency you can think of. Firefighters are called when there is a delay of EMS (emergency medical services). They must have highly diversified knowledge and skills in order to handle every type of emergency and non-emergency situation, with little or no warning or preparation; firefighters definitely fit the image of the word *hero*.

As well as rescues and medical emergencies, our days are filled with responsibilities that include maintenance of tools, equipment, and

Firefighters must adapt to any situation that arises. Firefighters are the end of the road. If firefighters can't handle a situation, then the situation can't be handled. They can't turn around and call someone else. It's critical that firefighters know their jobs and execute them without failure.

the station; pre-planning; and firefighters' assessment of local industrial and commercial businesses; community education; building inspections; personal development / training; and volunteering in the community.

Firefighters must be physically fit to meet the demands that arise under high heat conditions and other dangerous environments.

Firefighters have many roles and each one is equally important. Keep this in mind when you are selecting your courses and preparing your résumé.

Success seems to be largely a matter
of hanging on after others have let go.
— William Feather

ROLES AND RESPONSIBILITIES OF A FIREFIGHTER

A firefighter extinguishes fires, protects life and property, and operates and maintains equipment by performing the following duties:

- Creates and/or maintains access into structures
- Ventilates structures using natural, mechanical, vertical, or horizontal ventilation
- Responds to medical emergencies—chest pains, shortness of breath (SOB), stroke, seizures, diabetic reactions, shock, head injuries, etc.
- Restricts movement of a vehicle in a collision by using stabilizing technique
- Understands fire detection systems and alarm panels
- Maneuvers fire hose lines and handles nozzles, related hose line adapters, and various types of portable extinguishers
- Hooks hoses to hydrants to create a water source for fire suppression activities
- Understands sprinkler and standpipe systems, connections, and other special fire protection systems
- Maintains and operates illumination equipment, portable generators, and gas- and electric-powered fans
- Administers advanced first aid and performs artificial respiration and cardio-pulmonary resuscitation
- Operates mechanical resuscitators and defibrillation machines
- Enters burning structures to rescue and preserve the lives of occupants inside
- Properly uses extinguishers, axes, pike poles, hydraulic tools, pneumatic tools, cutting torches, and other equipment
- Uses various types of protective breathing apparatus when working in hazardous atmospheres
- Works off of ladders from various heights using an assortment of tools such as axes and pike poles
- Evaluates the placement of ladders for safe and effective usage
- Maintains and repairs firefighting equipment

- Understands and operates various types of detection instruments employed by the fire service to detect hazardous situations
- Becomes thoroughly familiar with city streets, fire hydrants, intersection building numbers, and building contents and occupancy
- Transporting firefighters safely to and from emergency scenes
- Operates aerial ladders, integrated or portable pumps to provide fire streams for elevated work, and rescue platforms
- Understands and recognizes fire behavior, such as back drafts, rollovers, and flashovers
- Continues learning in order to pass oral, practical, and written examinations mandated by department policy
- Remains current in fire safety practices and fire safety awareness with students and staff in public and local schools
- Responds to emergency call-backs while off duty
- Responds immediately and safely to all emergency calls and requests for assistance
- Maintains the physical and mental fitness necessary to carry out all duties of a firefighter
- Establishes and maintains the confidence of the public
- Attends training sessions
- Undergoes in-service inspections
- Pre-plans commercial and industrial establishments

CHAPTER TWO

HOW SOON SHOULD I START PLANNING?

High school is the ideal place to start planning and becoming enthusiastic about your future. It's where you begin to form the learning skills that will be needed to acquire an interesting career. It is also where you start to develop your life skills. Fire departments are interested in your entire background. If you have played many types of team sports, you will be seen as a team player who is able to take orders and understand the importance of working well with others. Your plans may be to become a firefighter down the road after you have obtained a university degree or acquired a skilled trade, but your activities in high school are nonetheless important. How much does it matter that in high school you were a volunteer in your community? A great deal! It shows the department that you have the caring attitude they're seeking. If you didn't volunteer in high school, it's not too late. Start now.

There aren't any particular high school courses to recommend. An early general education is good for most career choices. However, firefighting is a profession that requires diversified knowledge and skills, and you never know what skills, old or new, you're going to need and when.

Becoming a firefighter will not happen overnight. Succeeding in getting a job takes a great deal of patience and commitment. It may seem to you that you don't benefit a great deal from certain courses you take or things you do along the way,

View the process of becoming a firefighter as a job in itself.

but when you are ultimately hired by a fire department and receive the badge, everything comes into perspective. Every second spent to get there then seems worthwhile.

It can take a year or longer to acquire all the prerequisites and skills to become eligible for most fire department recruitments. It may take an extra year if you decide to enroll in a post-secondary firefighting education program. There are other fire-related courses available, and you may be prepared earlier if they are offered at times to suit you and if you schedule everything well. The sooner you start the better.

You have chosen a career that is highly competitive. It is important to have something to fall back on (a degree or a trade) in case you are not immediately hired by a fire department. Be wise—think ahead and make sure you have another job or career to maintain a satisfying and active life.

In order to discover new lands,
one must be willing to lose sight of the shore
for a very long time.
— Anonymous

CHAPTER THREE

IS AGE A FACTOR?

Is your age stopping you from pursuing a career in firefighting? Do you think you're too old or too young to get hired? These questions run through every potential firefighter's mind at some point. It's difficult to provide a straight answer to these questions because there are so many variables that affect the outcome. Fire departments may or may not have a specific age range from which they prefer to hire. Some fire departments recognize that some of the best candidates recruited fall under the so-called "too old or too young" category. The disadvantage of hiring younger firefighter recruits is their lack of life experience, but on the positive side, they have strength and endurance. If a mature firefighter recruit is hired, fire departments gain life experience and loyalty, but lose years of available service. It's stereotyping to assume that every younger or older recruit shares these qualities. There are certainly exceptions, but it's difficult for fire departments to assess candidates when departments are flooded with thousands of applications.

If you're a young firefighter recruit and you're worried that you might not be hired because of your age, here's a tip: Fire departments assume that, because of your age, you are lacking life experience and maturity, and that in most cases a young firefighter recruit is only interested in the thrill of fighting fires and saving lives. Can you imagine how impressed an interview panel would be if you showed an understanding of the importance of fire prevention? This would reflect your understanding of the various roles and responsibilities of a firefighter as well as demonstrate your maturity level.

How can your résumé reflect an interest in fire prevention? Simple—volunteer. Most fire departments have a fire prevention officer or include fire prevention duties. If you show an interest in accompanying a fire inspector to help out and to gain some experience, you may be allowed to ride along. Then, you will have related experience under the "Volunteer Experience" heading on your résumé. Another benefit is that the fire department will be able to put a face to your name, so the next time they're recruiting, it will be your name they recognize!

Fire departments may seem unpredictable in their hiring practices, and sometimes it's hard to figure out what their reasoning could have been for not hiring

you. It really depends on the structure of the fire department and its long-term goals. If you aren't sure whether the particular fire department hires older or younger firefighters, you can research the ages of the firefighters that have been hired over the past number of years and find out the average age. It is wise to refrain from disclosing your age during the hiring process, unless you are asked directly in an interview or on your application form. Employers must avoid questions regarding race, color, sex, religion, and national origin. It's also illegal to discriminate against someone because of age, so most fire departments don't want to know how old you are until after they've hired you. The bottom line is, you want to be someone who, because of your knowledge and skills, would be valuable to the department.

Start by doing what's necessary, then what's possible,
and suddenly you are doing the impossible.
— Francis of Assisi

CHAPTER FOUR

CAREER CHANGE

Sometimes the path you're taking isn't necessarily the path you'll end up on. Have you been thinking about a career change and considering the fire service as an option? This seems to be very common lately, perhaps due to an unstable economy and the fact that it seems more difficult to get work. Is now the time to jump ship and dive into the career you've always wanted—firefighting?

Either you've made the decision to leave your existing job or you've had the misfortune of losing your job. It doesn't matter now how you got here. All that matters is that you need to figure out how you can make this career change happen. It is normal to feel like you're starting over and that everything you've been working so hard for up till now—that is, your position and seniority—is or will be ancient history. You may also be looking at a pay cut, at least for the first couple of years upon being hired by a fire department, and that's if you're lucky enough to be hired right away. This can be a very stressful time, especially if you have lost your job and are forced into a career change. It can be very trying on your pocket-book, not to mention your family. So, if you're already at rock bottom as far as a job or career is concerned, then you have nothing to lose and everything to gain. You must believe in yourself and just go for it!

But firefighting isn't for everyone, and if at this point you've just been curious about the profession, then read on, because this book will paint a very clear picture of what you're up against, the level of commitment you will have to make, and what you can expect. So, at the very least, read this book from cover to cover, and then decide whether this is something you want to pursue.

If you have already made up your mind that this is what you want, then let's get started. I made a career change—after five years in the refrigeration trade, I decided that I wanted to become a firefighter. It was like one minute I had my whole life planned out, and the next minute I was heading down a path completely unfamiliar to me. Nevertheless, I just put my head down, did everything absolutely possible, and didn't look up until I was recruited. I believe wholeheartedly that you have to pursue this career with all you've got. There can be no doubt and there can be no halfway measures. You must be one hundred percent committed.

Now, the good news is that you're not starting from scratch. Your past experience is going to pay off for you in one way or another. Take me for example: I had my refrigeration license as well as my gas fitting ticket. I initially thought I was starting over, but that wasn't the case. I knew my skills would benefit me personally, but I hadn't really considered what the fire department would gain from them. It turns out that knowledge of these trades was extremely valuable to the fire service. Understanding aspects of how the world works can be to your advantage. If you have worked before and have been responsible for other employees, then you have an understanding of what an employer is looking for in an employee.

Is someone who has experience in the workforce, a family to feed, a good reputation, a clean medical and criminal record, and a lot of life experience to draw from, a better hire than a young, inexperienced recruit with little or no job history or references? Maybe yes, maybe no. It comes down to a matter of opinion. However, fire departments have been recruiting people with varied backgrounds for some time now. This career used to be a young man's game, when departments relied on strength and stamina to beat down fires. But these days, fire departments can have more when it comes to hiring recruits. Mature candidates, with both brains and brawn, seem to be leading the race in the recruitment pool these days.

CHAPTER FIVE

VISITING YOUR PHYSICIAN

Before you spend your time and money on firefighting school or fire-related courses, it is important to first visit your physician. Make arrangements to have your doctor conduct a thorough physical examination to confirm you are in good health. You must be able to meet the minimum physical health requirements of the fire departments. You must have a minimum of 20/30 vision in both eyes without the use of corrective lenses. However, corrected vision through laser surgery is acceptable. If laser surgery is not an option for you, then there is no point in spending a significant amount of time and money on school or courses. Also, be sure to confirm that you don't have any loss of color vision or hearing damage you were unaware of.

You should specifically inform your physician that you are thinking about a career in firefighting. This will ensure a thorough physical examination that is tailored to give you an accurate assessment of your physical condition. This way you will eliminate any surprises later in the recruitment process when you are undergoing a medical examination for the fire department. Be sure to discuss with your physician any medical concerns you may have; don't wait for a medical problem to be uncovered, especially allergies that could be corrected to increase your chances of being hired. For example, I played in a band for eight years and hearing loss was a concern of mine. So, before I took any other steps, I had an audiologist assess my hearing.

Firefighting is an excellent reason to kick poor lifestyle habits, such as smoking and excessive drinking. You'll not only feel better, but others will take you more seriously when you say that you gave up smoking to become a firefighter.

Fire departments expect you to be in excellent physical and mental condition and they don't accept anything less.

Because of the physical tests that are required and the physical strain endured on the job, it's imperative to be in good shape. It is wise to consult your doctor to make sure that your body and specifically your heart can handle the stress of working out and getting into shape. Your doctor will be able to help you find a safe target heart rate to use as a guideline while working out. If you have not trained in a while or you have never done so, be sure to take it easy at the gym for the first couple of weeks. Your body will need time to adjust to the new level of activity. However, your body should adapt quickly. If for any reason you feel light-headed or dizzy during or after a workout, be sure to consult your physician.

If you have any previous sports injuries or strains, take the time to review them with your physician or fitness instructor to develop exercises you can perform to prevent the strain or injury from reoccurring.

TIP!

Throughout your recruitment process and your firefighting career you are going to be demanding a great deal from your body. Be sure to take care of it.

"An ounce of prevention is worth a pound of cure."

CHAPTER SIX

THE RIGHT ATTITUDE

Do you have what it takes to be a firefighter? No one knows the answer as well as you. You're the one to decide your future. If you want to be a firefighter, the only one who can make that happen is you.

Attitude plays a major role throughout your recruitment process. I can guarantee that you're in for a lot of ups and downs. The ups are really up and unfortunately, the downs are really down. There is often no middle ground. It's important that you enjoy and hold on to the positive things when they occur to get you through the negative times. Becoming a firefighter is not an easy task. The profession is highly competitive. Don't let anyone discourage you from pursuing your dream; be true to yourself but don't think it's going to be a walk in the park. It's important that you understand and develop the commitment and determination required to be successful.

Try to maintain the same level of intensity throughout your recruitment process. By communicating with other candidates, you can help each other stay focused and informed. If you are taking a course, talk to the other students to find out what other courses they have taken. You may find out invaluable information regarding excellent courses to take—possibly ones you would have never thought of on your own. Or, you may decide not to take a course you had been considering because others didn't find it particularly helpful. You can't afford to waste time on your journey to your chosen career. A good way to join forums on the internet is through www.becomingafirefighter.com.

Preparing for the recruitment process takes a lot of time, concentration, and energy. You may have to miss important activities that conflict with commitments in your pursuit of a firefighting career. But that's the type of dedication this job demands—and deserves; each sacrifice, large or small, is worth it. If you asked fifty candidates whether there was anything they wouldn't do for a firefighting job, I guarantee you all fifty candidates would say "No!" So, you have to ask yourself that question, and if your answer is "Yes," then you should rethink your career choice, because those who have a greater commitment will have the greater edge. It's important that you realize the competition that you're up against, so be prepared to completely commit yourself to pursuing your goal. If you believe in yourself and never stop believing, your dream will come true. Be positive in everything you do.

Try to be pleasant and polite to everyone—firefighters deal with many different people in many different situations, including city hall workers, city maintenance workers, doctors and nursing staff, EMS personnel, police officers, charitable organizations, school staff and students, home and business owners during building inspections and the pre-planning of commercial establishments. It's important that you have excellent communication and interpersonal skills. Throughout your recruitment you're going to come into contact with many different individuals. With the proper attitude and the right approach, it's possible to gain supportive friends in high places. Being recruited into a fire department is a little easier if you have well-respected friends in your corner.

So spend your time gathering information, making the right kinds of friends and acquaintances, and taking courses. Knowledge is the key, so never stop educating yourself.

**Courage is resistance to fear,
mastery of fear-not absence of fear.**
— Mark Twain

ATTITUDE VERSUS QUALIFICATIONS

It's not whether you win or lose; it's how you play the game. How does this expression relate to pursuing a firefighting career? In my experience dealing with recruits across North America, I can definitely say that a good attitude is just as important as more easily measured qualifications.

Most often, what fire departments are seeking is a quality that can't be obtained by taking a course, or that doesn't expire like one of your licenses, but is chiseled deep into your soul—that is, a good attitude.

Attitude is what defines you as a person and as a recruit. "Attitude has a lot to do with interpersonal communications, self-esteem and your perceptions of others and theirs of you," Pam Wyess told participants in her Human Resource Development (HRD) class at the University of Michigan. "A good attitude is what gives leaders or instructors the sense that they can mold you into something better. A bad attitude is what curbs learning potential and limits opportunities."

Studies have shown that your attitude is communicated to others in three ways: by words, by tone of voice, and, most important, by non-verbal language, such as posture and facial expression. An interview is a good way of judging attitude, and interviewers' questions are often designed to assess your attitude. Don't be fooled into thinking you're being judged on the basis of coming up with the "correct" answers. You're also being judged on how you deliver those answers.

Have you ever wondered why some recruits walk away from interviews without a job offer, even though they have better qualifications than other candidates? A bad attitude may be the answer. Do your best to display a good attitude and chances are it will serve you well. Fire department recruiters have to have a sense that you're someone who can fit in well to the tight community within the fire hall. It's that whole "one bad apple" thing at work.

POSITIVE QUALITIES

Some of the best qualities a firefighter recruit can possess are:

- **Humility**—the ability to laugh at oneself

- **Modesty**—having or showing a moderate estimation of one's own talents, abilities, and value

- **Common sense**—sound judgment not based on specialized knowledge; native good judgment

- **Curiosity**—A desire to know or learn

- **Ambition**—An eager or strong desire to achieve something

The worst qualities to have in the absence of those listed above are pride (sense of one's own proper dignity or value) and overconfidence (a feeling of excessive assurance, especially of self-assurance).

In the absence of humility, modesty, common sense, curiosity, and ambition, pride and overconfidence come off as being cocky and arrogant. It's important to be able to laugh at yourself when you make a mistake and to back down when others may have another point of view. You should always listen to what others are saying because this is when you learn the most—not about others but about yourself.

Remember, nine times out of ten, the recruits that get hired are those who have both a good attitude and good qualifications. Fire departments are more likely to hire a candidate with the minimum qualifications coupled with a good attitude, over a highly qualified candidate with a bad attitude.

KEY POINTS TO REMEMBER

- There are benefits to being a young recruit as well as being an older recruit.
- Have a doctor assess your health.
- High school is a great place to start preparing yourself to be a firefighter.
- Never stop educating yourself.
- Attitude is everything!

NOTES

NOTES

Part II

Getting Ahead of Your Competition

CHAPTER SEVEN

MINIMUM REQUIREMENTS FOR BECOMING A FIREFIGHTER

Every fire department you approach will have a list of minimum requirements that you must achieve prior to applying for a firefighting position. These minimum requirements will vary from department to department and are based on the types of calls they respond to. A fire department located by water may require you to have taken a certain water rescue course, whereas a department located on an escarpment may require you to have your high angle rescue technician level. Be thorough when researching the minimum requirements for the fire departments that interest you most. Many are similar, so you can obtain qualifications or be eligible to work at a number of different departments. You should be able to determine the specific requirements by either contacting the city's human resources department or by visiting each fire department's website.

Always be prepared. Don't wait for the fire department to post a recruitment, only to find that you're not eligible to become a recruit. This would be especially disappointing if the recruitment was in your home town. It is important to ensure that the information you have is accurate and up to date.

Take a moment and brainstorm what you think the minimum requirements of a firefighter candidate should be. Put yourself in the fire department's shoes, and let's see what you come up with. Use Figure 7.1 for your brainstorming. When you have completed your list, compare your answers with an actual fire department's minimum requirements listed on the next page.

FIGURE 7.1

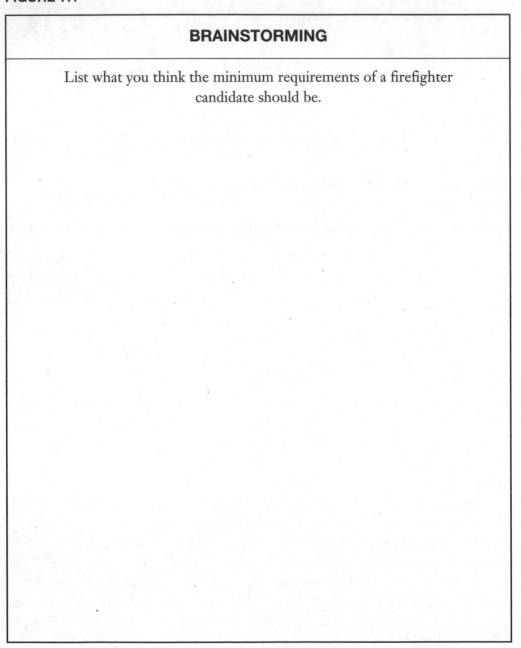

BRAINSTORMING

List what you think the minimum requirements of a firefighter candidate should be.

EXAMPLE OF MINIMUM REQUIREMENTS:

- Be legally entitled to work where you are living as a citizen or landed immigrant.
- Provide proof of Grade 12 education or equivalent.
- Meet the prescribed visual requirements of 20/30 in each eye without the use of corrective lenses and meet a color vision assessment.
- Be able to communicate effectively in English.
- Provide a copy of a valid CPR and Standard First Aid Certificate Level C.
- Provide a copy of a valid unrestricted Class 3 or Class D driver's license with an air brake endorsement (or Z endorsement).
- Provide a copy of a valid certification of successful completion of the firefighting physical fitness test.
- Successfully complete an aptitude test and associated interview.

Although the minimum qualifications are fairly easy to achieve, the fact is that firefighter candidates raise the bar each time they apply. Many candidates have taken lots of courses and achieved greater skills to improve their chances of success. In order to stay competitive, you should also achieve more than the minimum requirements. If you have the right attitude, you should not be discouraged; just accept it as giving you an advantage. Don't waste time—get going on your plan to take the courses that you determine will best fill the gaps in your résumé or add strength to it. It is important to sit back and carefully assess what you need to take. For example, there is no sense in taking a course that is similar to one you already took if you haven't obtained all of your prerequisites or if other courses will give you a broader education. Plan carefully to keep you on the right track.

CPR Level C
covers all aspects of
CPR skills and the
theory for adults,
children and infants.

ACHIEVING MINIMUM REQUIREMENTS

Achieve the minimum requirements for your first choice of fire department before achieving them for your second choice or third, especially if you know your first choice is going to be recruiting in the near future. But you must also use common sense. Obviously, if your third choice is having a recruitment in the near future and your first and second choices have no intentions of having a recruitment in the foreseeable future, then obtain all of the third choice minimum requirements first.

But be realistic. Even though you may be focusing on three departments, you still have to be willing to get hired anywhere.

FIGURE 7.2

What are the three fire departments that you would like to be hired by?
First Choice _____
Second Choice _____
Third Choice _____

It's impractical to have the minimum requirements for every fire department in your region. However, you may find the minimum requirements for one department are similar to others. Keep a good written record of your research. You can use the organizational chart in Figure 7.3.

FIGURE 7.3

**What are the minimum requirements for each
of the fire departments listed on page 28?**

First Choice	Second Choice	Third Choice

Now, use a chart like the one in Figure 7.4 to help you figure out how to obtain these minimum requirements. List all the minimum qualifications you need to obtain for your first three choices. Then, do research on each qualification to determine what it will take to achieve it, how long it might take, when the course is offered, and the locations that offer the qualification.

FIGURE 7.4

FIREFIGHTER QUALIFICATION ACTION PLAN		
Minimum Qualifications	**What Will It Take? How Long Will It Take?**	**Where and When Is It Offered?**
Example: - *Class C or Class D Driver's License with Air Brake Endorsement Z*	*Example:* - *Obtain Beginners License* - *Sign up for and complete driver's course* - *Obtain medical exam* - *Pass driver's test* - *approx. 1 month*	*Example:* - *Ontario Truck Training Academy, transportation office (on Exeter Rd.)* - *offered as needed/anytime*

RECOMMENDATIONS

It is also very important at this early stage to start learning about other aspects of the fire departments you are pursuing. Although volunteer experience isn't mandatory, it is considered a major factor in the hiring process. Try to find out which charities they support. This can help you decide where to start volunteering and making connections. If one fire department raises money for children with learning disabilities, wouldn't it make sense for you to volunteer with kids who have learning disabilities? Volunteering will be discussed in more detail in Chapter 10.

BRANDON MEETS HIS HERO AT STATION 245

When 6-year-old Brandon Quesnelle pulled a pot of boiling oil on top of himself in a cooking accident almost two months ago, it was Toronto firefighters who tended to his injuries and started him on the long road to recovery.

Yesterday the youngster re-established the bond that was forged in that moment of intense pain and fear, visiting the firefighters at Scarborough Station 245 at Birchmount Rd. and Ellesmere Rd.

The blue-eyed boy in a blue polo shirt stepped out of his mother's van at the station.

He was greeted by firefighter Brian Fogarty, who treated him in the ambulance after the scalding oil ran down his shoulder and back, causing severe burns. Brandon spent the next five weeks in hospital and was only able to return to school on March 27.

"Do you remember the ambulance ride?" Fogarty said.

"No," said Brandon, who then sat on the ground and crossed his legs. "Just my leg hurts when I do this."

Then the little boy wanted to see the fire truck and slide down the pole and try on their boots.

His eyes lit up when he saw the presents from his heroes sitting on the bumper of the truck. The package include a Tonka fire engine, a hockey stick and a Scottish national soccer team jersey.

He also received a package from Camp BUCKO—Burn Camp for Kids in Ontario—sponsored by the Toronto Firefighters Association. Brandon can go for one week each summer for 10 years.

"We don't really get many good moments," Fogarty said after the meeting. "He looks a lot better than I had expected."

— Himani Ediriweera. *Toronto Sun*. April 3, 2006

CHAPTER EIGHT

CHOOSING THE RIGHT COURSES

Through diversified training and continuing education, firefighters have equipped themselves to handle any type of situation. No matter what the call of duty—a motor vehicle collision, a medical emergency, a hazardous materials incident, or a fire, firefighters are ready and able to handle it. Firefighters also need to know how to maintain and use stand pipe and sprinkler systems and must have an in-depth knowledge of building construction. In order for you to be well qualified, you will have to spend a lot of time and a substantial amount of money. If you're just starting to pursue a career in firefighting, you will face competition from other candidates who have already done the same.

So how are you supposed to measure up to becoming a serious candidate? It really depends on how motivated you are as well as the amount of time you can dedicate to this process. In terms of choosing courses, I would advise against choosing what I call "cookie cutter" courses—those that most everyone has acquired — such as high angle rescue awareness level, the most basic rope rescue course available. To clarify, high angle rescue is a valuable course to have, however a very common qualification to obtain. You need to be different and take courses that will set you apart from everyone else. You want the recruiter to be impressed with the thought you put into it and how

Offer the fire departments your full potential. Help them create depth in their fire personnel.

hard you have worked to acquire the courses listed on your résumé.

Too many candidates say to themselves, "If I'm lucky, I'll get that first interview." Luck has nothing to do with it! Whenever I asked myself, "Would I hire me?" I didn't stop taking courses and improving myself until I could honestly answer "Yes."

COURSES TO CONSIDER

An easy course to obtain in a weekend that is often overlooked by candidates is the certification to install car seats. Many fire departments are involved in car seat installation clinics that help the public install their car seats properly. These courses are usually available through city's health units, if not contact your local fire department and see if they can help you locate a course provider.

Fire Training
- EST (Emergency Services Technician) program
- Pre-service firefighter certificate
- Scuba license
- Pump "B" operator
- Silo fire awareness
- Fire prevention / investigation division
- Fire arson investigation
- Air pack maintenance, field level
- Marine firefighting for land-based firefighting
- Officer 1021
- NFPA Courses:

NFPA 1000
> *Standard for Fire Service Professional Qualifications Accreditation and Certification Systems*

NFPA 1001
> *Standard for Fire Fighter Professional Qualifications*

NFPA 1002
> *Standard for Fire Apparatus Driver/Operator Professional Qualifications*

NFPA 1003
> *Standard for Airport Fire Fighter Professional Qualifications*

NFPA 1005
> *Standard on Professional Qualifications for Marine Fire Fighting for Land-Based Fire Fighters*

NFPA 1006
  Standard for Rescue Technician Professional Qualifications
NFPA 1021
  Standard for Fire Officer Professional Qualifications
NFPA 1031
  Standard for Professional Qualifications for Fire Inspector and Plan Examiner
NFPA 1033
  Standard for Professional Qualifications for Fire Investigator
NFPA 1035
  Standard for Professional Qualifications for Public Fire and Life Safety Educator
NFPA 1037
  Standard for Professional Qualifications for Fire Marshals
NFPA 1041
  Standard for Fire Service Instructor Professional Qualifications
NFPA 1051
  Standard for Wildland Fire Fighter Professional Qualifications
NFPA 1061
  Standard for Professional Qualifications for Public Safety Telecommunicator
NFPA 1071
  Standard for Emergency Vehicle Technician Professional Qualifications
NFPA 1081
  Standard for Industrial Fire Brigade Member Professional Qualifications

Rescue

- Bronze Cross
- Bronze Medallion
- Confined space rescue
- Farm accident rescue
- Hazardous materials awareness, operations or technician level
- Ice rescue
- National life guard qualification
- High angle rescue
- River rescue
- Swift water rescue

Medical
- AED (Automated External Defibrillator)
- Air ambulance landing zone preparation
- BTLS (Basic Trauma Life Support)
- EMR (Emergency Medical Responder)
- First aid instructor
- Intermediate cardiac life support
- Paramedic
- Pediatric Pre-hospital Care

Operators
- Aerial work platforms
- Class 1 or class A license
- Class 2 or class B license
- Class 3 or class D license
- Class C license
- Defensive driving
- Industrial truck operator
- International hand signs (rigging)
- Pleasure craft operator license

Communication
- Sign language
- Emergency services communicator certificate
- Public speaking
- Crisis intervention
- Emergency medical dispatcher
- Restricted radio operations license

General
- Business administration courses
- Any trade: electrician, plumber, refrigeration
- Blueprint reading
- Building construction
- Carbon monoxide

- Car seat installation (baby seats)
- Critical incident stress management
- Energy control and power lockout
- Fall arrest certificate
- Fire alarms
- Gas technician I, II, or III
- Health and safety
- Outdoors recreational leadership diploma
- Service and repair of small engines
- Sprinkler systems
- Transportation of dangerous goods

This list may seem overwhelming. The best way to decide which courses to enroll in is simply to take the same approach discussed for obtaining your minimum requirements. Focus on three fire departments and make sure you take the courses that will benefit the fire department where you first wish to be hired. Then, take other courses that would be good to have (see the other departments' requirements). For the three fire departments you choose, find out specifically the kinds of emergency calls that each department commonly responds to. If one of the three requires high-angle rescue emergencies but all three do water rescue emergencies, it would only make sense to get your water rescue certification before the high-angle rescue. Prioritize courses to make sure you first obtain those that are prerequisites for the departments you are interested in. Your main concern is to be eligible when fire department recruitments occur. You may well be hired before you spend a lot of money on non-prerequisite courses—but keep taking extra courses to maintain your edge.

SCENARIO – WHICH COURSES TO TAKE?

Recently, I had a young man visit me at the fire hall. I had never met him before, and he had just found out about my service and my involvement with firefighter recruits. He worked at one of the local car dealerships as a mechanic. His question was, "What courses should I be taking to improve my chances of being hired as a firefighter." My first question is always, "Do you have your NFPA 1001 or your Pre-Service Firefighter certification (recommended qualifications for Ontario)? His reply was that he had already obtained his NFPA 1001 as well as all the other minimum requirements for Ontario, (DZ License, CPR, first aid, etc).

His seven years of experience working as a mechanic told me he is good working with his hands, is employable, is reliable, understands a hard day's work, and is able to learn new skills (because, as we all know, the car industry changes constantly when it comes to technology). So, my second question to him was "What are your hobbies—what do you like to do?"

"I love cars," he replied, and I said, "That's perfect!" He looked at me as if I had two heads. But my suggestion was that he start to combine his love and knowledge of cars with skills that would be useful as a firefighter. Fire departments place a high value on having someone who is educated and skilled in one specific aspect of firefighting. I suggested he focus on vehicle extrication—become the go-to guy when it comes to cutting edge technology in vehicles such as hybrid cars, safety systems such as air bags, seat belt pre-tensioners, ROPs (rollover protection devices), and alternate fuel sources such as electric and hydrogen. His eyes lit up, and I could tell this was something he was very interested in. He told me he had already given the fire department a crash course on hybrid cars.

This recruit has since taken every available course offered by his car dealership regarding all new car technology (and it didn't cost him a cent!) and has enrolled in several vehicle extrication courses. So, he has both improved his skills in his current job and improved his chances as a firefighter recruit.

COURSE BRAINSTORMING

Throughout your recruitment process you will find that you require more courses to remain competitive. You may find that someone recommends a course, but by the time you get to the point where you seriously consider taking it, you've forgotten what exactly it was or where it was offered. There are so many courses, locations, and dates to remember that it is impossible to keep track of everything.

Whenever you come across a course or certification that may be valuable, write it down immediately as part of future planning. The course brainstorming chart in Figure 8.1 will allow you to record courses and certifications you come across and to research them when you have spare time. Once you have done your research, transfer them into the "Short Courses" or "Long Courses" column of the chart.

Short Courses. These are courses that you can take with limited time and money. Usually these are one-day or weekend courses which are offered frequently.

Long Courses. These are courses that are more involved and usually require a greater commitment. More planning is required.

Choose a course from your short courses list when time and money are limited. If you anticipate an increase in available time in the near future, you may consider taking a course from your long courses list. By taking this approach, you are always focused and ready to obtain qualifications that are going to place you ahead of your competition.

People with goals succeed
because they know where they're going.
— Earl Nightingale

FIGURE 8.1

COURSE BRAINSTORM

1.	15.
2.	16.
3.	17.
4.	18.
5.	19.
6.	20.
7.	21.
8.	22.
9.	23.
10.	24.
11.	25.
12.	26.
13.	27.
14.	28.

SHORT TERM COURSES   LONG TERM COURSES

1.	1.
2.	2.
3.	3.
4.	4.
5.	5.
6.	6.
7.	7.
8.	8.
9.	9.
10.	10.
11.	11.
12.	12.
13.	13.
14.	14.

You may also find the following organizing tool helpful. The Course Manager in Figure 8.2 helps you to keep a log of courses you have taken, courses you have signed up for, and courses you plan to take in the future.

COURSE MANAGER

Take advantage of this organizing tool. The "Course Manager" chart allows you to keep a log of the past and upcoming courses taken or you want to take.

FIGURE 8.2

Course	Date	Length of Course	Location	Date Course Completed	Date Added to Résumé

There are many different ways to get a job at a fire department and it is important to figure out which path is the best for you. For example, a first aid and CPR instructor's certificate would be an excellent qualification to have on any résumé. However, you may wish to postpone these in favor of other courses you are interested in and believe would benefit you more specifically with the departments you are interested in.

Just remember to make every minute of every day count. Do something every day to keep you on track, whether it's visiting a fire hall or reading a newspaper article about a recent fire in your hometown—even cut out articles and start a scrapbook. There are many ways to keep your head in the game. Find out what works best for you.

Goals are dreams with deadlines.

— Diana Scharf Hunt

MEDICAL TRAINING

Times have changed. Firefighters respond to many different emergency situations, and not only to fires. In fact, the majority of emergency calls are medical-related emergencies. Most fire departments are on the scene within four to five minutes of receiving the alarm, and often before an ambulance. In many cases, those few minutes can be the difference between life and death.

Before you spend a lot of time, money, and energy in taking medical courses, be sure to research which regions recognize that particular type of training. Each region has different policies and guidelines regarding the minimum requirements for each level of emergency medical certification. For example, Ontario doesn't recognize EMT-B (Emergency Medical Training-Basic), but for most provinces in Canada, EMT-B is a prerequisite for any candidate applying for a firefighting position. Be sure to pay close attention to these requirements because fire department protocols change often.

If you've already obtained a medical course certificate that isn't recognized in a specific region, chances are you will still get credit for the training even though you won't be able to use it. This means that you can't perform to the level in which you've been trained, so in this case the course may have got you the job even though you can't completely use your skills.

If you're applying to a fire department that only requires you to possess a minimum of basic first aid and CPR—Level C, then any medical training you receive beyond that will be a bonus. Unfortunately though, for those fire departments that recommend higher levels of medical training such as EMT-B or Paramedic, you may have to find other areas on your résumé to pick up bonus points.

However, no matter what level of medical training you may have, it is the candidates with actual experience who will have an advantage.

FIGURE 8.1

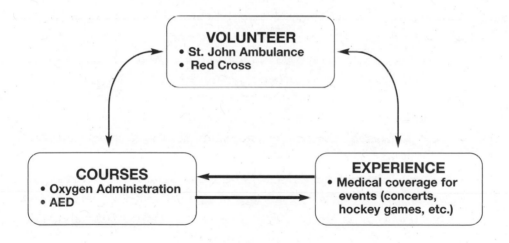

My advice is to start volunteering with a medical response or medical-related non-profit organization. You will accomplish two objectives simultaneously: building volunteer experience and gaining confidence handling medical non-emergency and emergency situations. Having this experience looks good on your résumé and will assist you in your interview when asked questions such as, "Was there a time when you had to respond quickly and effectively to an emergency situation?" Being able to give a detailed answer (i.e., date, time, place) will enhance the impression you make at your interview.

FIREFIGHTER GOES BEYOND THE CALL OF DUTY

Saving lives is in a firefighter's job description. Seth Wells figured he would just go about it in a different way.

The 27-year-old civilian firefighter at Vandenberg Air Force Base returned home from Boston one weekend after donating his kidney to a fellow firefighter he had briefly met. Wells gave a kidney to Walter Stecchi, 45, who worked at Vandenberg in the 1980s and is currently an assistant chief at Otis Air Force Base, Mass.

Stecchi's kidneys had been going downhill for nearly 20 years and his body had already rejected a kidney from his father. Stecchi was on dialysis when his firefighters union sent a letter to other bases requesting donors. Wells immediately stepped forward, something Stecchi wasn't expecting.

"I was happy, surprised. I thought it was pretty amazing someone would step up like that. I didn't have much hope in the beginning," Stecchi said. Wells first met Stecchi when he flew to Boston in September to undergo testing. He met him a second time last week, before surgeons performed the surgery.

The transplant was an immediate success and the effect on Stecchi was like night and day, according to both men. After leaving the hospital, Wells and his family spent the week at Stecchi's home before returning to the Central Coast Friday night. Wells' colleagues from Vandenberg Fire Department gathered at the airport in Santa Barbara to greet him, his wife and child upon their return.

Wells is eligible for 30 days off and says Vandenberg officials have told him to take as much time as he needs. Stecchi's insurance paid the hospital bills and Well's firefighter union paid for his travel expenses.

"A lot of guys think I'm nuts. It's not about what if I need the kidney. It's what if I don't need it?" Wells said from his Grover Beach home where he is recuperating. "We're in this career field where we're trying to help people so I try to do that to the fullest extent."

Stecchi says he considers Wells a brother and both plan to stay in touch. "It's nice to have another set of family," Wells said. "We're going to meet every year, just hang out. It'll be a bond for many years to come."

— Mark Baylis, www.firerescue1.com

CHAPTER NINE

CHOOSING A FIREFIGHTING PROGRAM

You want to attend fire school, but which program should you take? And which school should you attend? Finding the answers can be frustrating. Ask ten different people and you will get ten different answers—and everyone would be right in some way. Choosing the right program for you will depend on a number of factors—geography is one. Generally, fire departments endorse or are loyal to schools that are in the same province or state. They have a thorough understanding of the curriculum and are often familiar with the instructors. In some cases, the local fire department plays a role in the instruction, and this actually allows the department to prescreen firefighting candidates and recruit the best in the class.

Other factors include your level of commitment, your family obligations, and your financial situation. If you're someone who has a young family and is paying down a mortgage, then it may not be practical to enroll in a three-year, full-time post-secondary fire program. You may have to consider taking a condensed program or distance-education program, or completing your course work over an extended period of time.

On the other hand, if you're someone fresh out of high school, then taking a longer post-secondary fire program will allow time for you to mature and gain valuable life experience. Both of these credentials will benefit you down the road. Either way, you'll have to make sacrifices, both big and small, but the commitment to training will help you invaluably in widening the gap between you and your competition.

When choosing a fire school or program, you must ask these questions: Does the geographic area you want to work in recognize the program? In other words does the program meet the required standards? The last thing you want is to spend your time and money on a fire program that isn't recognized or accredited. Governments and organizations regulate the standards of fire training and accredit schools to teach to that standard. These regulations are in place to ensure that you receive a level of training that gives you a strong foundation in which to start your firefighting career.

FIRE SERVICE ORGANIZATIONS

There are three fire service organizations you should be familiar with. It is important that you understand their significance and the relationship between them when choosing a fire program.

The **National Fire Protection Association (NFPA) www.NFPA.org** is an international non-profit organization founded in 1896. NFPA serves as the world's leading advocate of fire prevention and is an authoritative source on public safety. NFPA's three hundred codes and standards influence every building, process, service, design, and installation, including those in many countries. There are more than six thousand volunteers from diverse professional backgrounds that serve on 230 technical code and standard-development committees.

The **International Fire Service Accreditation Congress (IFSAC) www.IFSAC.org** is a peer-driven, self-governing system that accredits both fire service certification programs and higher education fire-related degree programs. IFSAC is a non-profit organization authorized by the Board of Regents of Oklahoma State University as a part of the College of Engineering, Architecture and Technology. The IFSAC administrative offices are located on the Oklahoma State University campus in Stillwater, Oklahoma.

The **International Fire Service Training Association (IFSTA) www.IFSTA.org** was established in 1934 as a non-profit educational association for the training of firefighting techniques and safety. IFSTA's purpose is to approve training materials for publication, develop training materials for publication, edit proposed rough drafts, add new techniques and developments, and delete obsolete and outmoded methods. It operates in partnership with Fire Protection Publications, whose primary function is to publish and disseminate training texts as proposed and validated by IFSTA. As a secondary function, Fire Protection Publications researches, acquires, produces, and markets high-quality learning and teaching aids that are consistent with IFSTA's mission.

To summarize, the NFPA develops the standards, IFSTA develops training materials based on the NFPA standards, and IFSAC accredits those organizations that certify firefighters based on NFPA standards as well as those institutions that offer fire-related degree programs.

Motivation is what gets you started.
Habit is what keeps you going.
— Jim Ryun

However, Ontario, has its own standard of firefighting training. The **Office of the Fire Marshall** (OFM) **www.ofm.gov.on.ca** has created its own curriculum that is recognized by fire departments only in Ontario. The OFM operates under the Ministry of Community Safety & Correctional Services. The role of the OFM is to minimize the loss of life and property from fire in Ontario by providing support to municipalities and fire departments across Ontario in meeting the needs of their communities, including public education, fire prevention, firefighting, fire protection, training, and fire investigation; providing leadership within the Ontario government by advising on standards and legislation relating to fire prevention and protection; and making recommendations for the provision of adequate levels of fire safety for buildings and premises within Ontario.

So, make sure that, when choosing a program, you enroll in one that is either accredited by the NFPA or government-approved in your province or state. Always check with the fire departments you are interested in.

There is one more factor when choosing a school. You must decide the type of firefighter you want to be, for example, structural, wildland, industrial. You will then need to research the schools that offer programs in that specific field.

To further help you assess whether the program you are planning to enroll in is the right one for you, consider talking to graduates of the program. Ask what they thought of their experiences, the curriculum, and the expertise of the school's instructors.

DISTANCE LEARNING

A fairly new option is distance learning, or online training, and is quite appealing to those with busy schedules or those who are not located near a school. The completion time for these programs vary. Some allow you to work at your own pace, while others are structured so that you have a written test every week on a specific day, at a specific time. Once you have completed the written portion of the training, you will then have to attend a fire academy for a condensed period of time to complete the practical portion of the program.

If this appeals to you, I recommend you still do research to ensure that the fire departments you are most interested in accept this type of training and certification. Many successful recruits have taken the online training programs, but this type of training is still new to the fire service, and some fire departments may be skeptical of this style of training. Again, always do your research.

FIGURE 9.1
RELATIONSHIP BETWEEN FIRE SERVICES ORGANIZATIONS

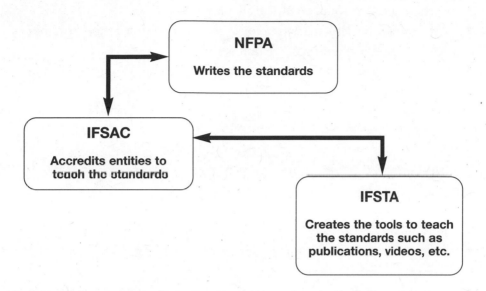

QUESTIONS TO ASK

Criteria for Acceptance

- Will I have to undergo a physical fitness test? Which one?
- What level of education is required—Grade 12 diploma? Chemistry? Biology?
- What is the language of instruction?
- Are high school transcripts required?

Information about the School

- Is the school accredited or approved?
- Does the province or state you want to work in recognize the school's program?
- What qualifications will you receive upon completion of the program?
- How many students are in each class?
- How much and what type of practical training is given?

- Is the school setting relaxed or highly structured like the military?
- Is extra help available (like mentoring) if I need it?
- Is there a recreational facility available?
- Are courses available part-time or through correspondence / online?
- Does the curriculum meet current standards?
- Is the equipment (e.g., training props, apparatus) in good condition and relevant?

What are the Costs of:

- Tuition
- Books and supplies
- Residence
- Parking
- Meals—cafeteria, restaurants, personal fridge space, etc.
- Room and board (if available)
- Transportation—taxis, buses, etc.

Researching and recording answers to these questions will make your decision easier. Use the chart in Figure 9.2 to help you organize your information.

FIGURE 9.2 • SCHOOL SELECTION ORGANIZER

	First Choice	Second Choice	Third Choice
School			
Program			
Address			
Website			
Email			
Phone			
Fax			
Length of program			
Part-time or distance learning			
Certification obtained			
Dates of program			
Cost of tuition			
Living costs			
Accreditation			
Approval by recruited fire departments			
Length of waiting list			
Prerequisites for admission			

CHAPTER TEN

VOLUNTEERING

How much does volunteering matter on your résumé? How much volunteering is enough?

These are two very good questions with no clear answers. I believe that without any volunteer experience, you will be at a disadvantage. Many fire departments look at volunteering as an indication of how much you care about your community. Since fire departments are very active in their community, they like to see the same from their potential recruits. Volunteering demonstrates positive characteristics that are valued in a firefighter, and I recommend you take it seriously. It's hard to say exactly how much of your time you need to dedicate, but make sure you can demonstrate that you have made your community part of your life.

Where you decide to volunteer is completely up to you, but your efforts should be directed toward an experience that demonstrates that you have an ongoing commitment to your city or town. So, volunteering for two hours at a food drive one day last year won't compare favorably to someone who has volunteered two hours a month or a week for the last year or two at the local hospital. Be aware that most hospitals have volunteer opportunities available, and if it is possible get involved in the emergency department, and so receive invaluable hands-on medical experience, that could give you an edge.

Volunteering is also a great way to meet other individuals who may also be pursuing firefighting as a career. It's important to interact with these individuals because you can help each other. It's difficult to become familiar with all courses being offered and everything that's happening regarding local recruitments. Friends with the same or similar goals can help you know what's going on.

There may also be monetary advantages to volunteering that you may not have considered. Some volunteer organizations require their volunteers to have specific training. Chances are the training will be supplied free of charge or at a reduced rate. There are three benefits: you're gaining volunteer experience, you're obtaining courses free or at a reduced cost, and, of course, you're helping your community.

When you are researching the top three fire departments of choice, inquire about the charitable or community organizations in which they are involved. If the chief of one of your fire departments is involved with an organization that helps children with autism, you can't go wrong by volunteering for an organization with similar goals.

Another excellent way to gain invaluable volunteer experience is to assist local fire departments with their own fundraising events. It takes a number of volunteers to run and organize a successful event, so fire departments usually appreciate any help being offered. What a great way to get your foot in the door and put a face to your name.

High achievement always takes place in the framework of high expectation.
— Jack Kinder

! KEY POINTS TO REMEMBER

- Minimum requirements may vary from fire department to fire department.
- Choose courses that set you apart from your competition.
- When choosing a firefighting program, be sure to choose one that is recognized by the fire department where you wish to get hired.
- Volunteer experience is equally as important as courses on your résumé.

NOTES

_____

_____

_____

_____

_____

_____

_____

_____

_____

_____

_____

_____

_____

_____

NOTES

NOTES

Part III

Recruitment Process Strategies

CHAPTER ELEVEN

MUNICIPAL FIREFIGHTER RECRUITMENT PROCESS

Fire department recruitment processes vary from department to department. Each department has the authority to establish the measures to be taken and the minimum qualifications and standards required to find the best applicants. Recruitment processes can be long and complicated or short and simple. The objec-

tive of a fire department recruitment process is to determine an applicant's aptitude as a firefighter. Fire departments use multiple methods to evaluate each applicant's ability to execute firefighting-related tasks, which includes assessing the overall health and fitness level of each applicant; the ability to function under physical and mental stress, and the ability to understand written and verbal communications.

Availability of fire department positions is unpredictable. It is difficult to determine when a fire department will announce a recruitment. You may hear rumors that a fire department is hiring long before a recruitment is posted. To be a successful recruit, you have to be prepared at all times. Sometimes there are multiple fire departments recruiting at the same time and sometimes there are none. You must be patient and maintain a positive attitude while waiting for opportunities to arise.

STAGES OF RECRUITMENT PROGRAMS

Fire departments use recruitment programs to help them find the best candidates for the job. The following list outlines the common stages of a recruitment program.

- Posting of the recruitment
- Application submission deadline
- Assessing whether the applicant meets the minimum qualifications
- Written aptitude test
- Notifying the applicant of the test results
- Scoring applications
- Notifying the applicant of the first interview
- Notifying the applicant of the second interview
- Arranging for a physical fitness test
- Arranging for a medical examination
- Confirming the applicant passes both the physical ability and the medical tests
- Contacting references
- Making a job offer

If you have been informed that a fire department is recruiting, be sure to find and carefully read the posting yourself.

The following is an example of a recruitment posting:

PROBATIONARY FIREFIGHTER RECRUITMENT   July 18, 2006

The Winnipeg Fire Paramedic Services is conducting a recruitment process effective 5 November 2005, with a closing deadline of 4:30 p.m. on 19 December 2005. The PDF document entitled Consolidated Recruitment Documents contains all the printable forms and requirements that candidates are required to fill out or provide. Please print and complete them exactly as indicated. WFPS thanks all applicants, but only those whose applications were deemed suitable for progressing to the testing and interview stages will be contacted.

QUALIFICATIONS:

1. Grade 12, according to Canadian Provincial Standards, GED or equivalent
2. CMA Accredited Paramedic Program (minimum PCP)
3. Possess a valid Provincial License or be able to acquire
4. IFSAC or Pro board Accredited NFPA 1001 Firefighter Level II program
5. Current CPR Certification
6. Clear Criminal Record check
7. Valid Class 4 Driver's License or equivalent with no more than 4 demerits and no alcohol related charges / convictions for the last 4 years and a Drivers Abstract
8. CPAT Certification (Candidate Physical Ability Test) or a certified job related physical fitness test acceptable to the Service completed within the past 12 months
9. The ability to successfully complete the EMS and fire Rescue Aptitude assessments, EMS and fire scenarios and a driving evaluation
10. Ability to undergo and pass a medical examination

Submit to: Human Resources
Fire Paramedic Service
2nd Floor, 185 King Street
Winnipeg, Manitoba R3B 1J1
(204) 986-6369 www.winnipeg.ca
(204) 986-6369 www.winnipeg.ca

Meeting the Application Submission Deadline

Once the decision is made by the municipality to recruit additional firefighters, the first step in the process is posting a public announcement. This is accomplished by placing an advertisement in or sending a press release to the media, such as local newspapers and the Internet, on the fire department's website, or on the city's website located under "Employment." Candidates must pay close attention to the details in the announcement, including where to obtain an application form, the application deadline, and where the application is to be submitted. Never trust second-hand information. Always confirm the recruitment particulars by examining the posting yourself.

It goes without saying that you must meet the application submission deadline. No exceptions are made for late submissions or failure to pay the application fee. Do not submit your application at the last minute—it looks like you may not be on top of things. Failure to submit on time convinces the fire department that you cannot follow instructions and that you lack self discipline. Note: You should notify the human resources department or the fire department in writing of any additional skills and education acquired since your application was submitted. Copies of certificates, diplomas, and so on should accompany the notification.

Checking Minimum Qualifications

Fire departments will confirm every qualification reported on your résumé.

Written Aptitude Test

In order to advance in the recruitment process, applicants must successfully pass the written aptitude test presented by the fire department. Fire departments rely heavily on these tests as a way to ensure the quality of the applicants.

Notifying the Applicant of the Test Results

Applicants can expect to receive their test results within a relatively short time after completion, though a large number of applicants may mean a longer wait. Typically, test results are mailed to applicants, and you will be notified of whether you have advanced to the next stage in the recruitment process. Unsuccessful applicants will receive a letter thanking them for their interest in the fire department and advising them that they will not be able to continue in the recruitment process.

Scoring Applications

Résumés will be assessed at this time to determine which applicants are worthy of receiving an interview. It is as this stage that your completion of courses or programs and your volunteer efforts will pay off.

Background Check

After the fire department is satisfied that you have met the minimum requirements, it will run a background test on you. There are ways to ensure that your background check will be as positive as possible. Take care of things that might draw attention to the background investigator.

- Pay your old traffic tickets.
- Obtain proof of full recovery after a serious illness.
- Ensure you are drug free.

First say to yourself what you would be;
and then do what you have to do.
— Epictetus

The First Interview

Failure to meet any one of the minimum requirements means your application will be moved to a "failed" file.

The purpose of an interview is not to explore the applicant's technical knowledge, but to determine further the applicant's employment experience, related skills, and volunteer experience. It will also allow the interview panel or hiring committee an opportunity to assess the applicants' personality and interpersonal skills. The applicant will likely be asked why he or she wants to be a firefighter and why the applicant feels he or she should be considered over other qualified applicants.

The Second Interview

Fire departments may use two or more interviews to find the ideal candidate. This is more common when the number of applicants is high. Fire departments typically notify successful first-round interview applicants of a second interview by telephone. The date, time, and location of the interview will be provided at this time.

Physical Fitness Test

Once the applicant has successfully passed the interview stage, the next hurdle is the physical fitness test.

Medical Examination

It is mandatory to have a thorough medical examination completed prior to being offered employment by the fire department. You should have discussed with your doctor any health concerns or issues prior to this time in order to be in good health for this examination.

Confirming Applicant Passes Both the Physical Ability and Medical Tests

Fire departments must have written documentation of both physical and medical assessments to confirm you are fit for duty.

References Contacted

A complete background search and reference check is completed by the recruiting fire department. It is in the best interests of the applicant to disclose any information that may be of concern to the fire department; honesty is vital to your success.

Applicants failing to meet any of the following requirements may be disqualified from the recruitment.

- Failure to submit recruitment entrance fee
- Failure to submit a fully completed application form
- Failure to submit documentation supporting your qualifications on your résumé
- Failure to submit required documentation (driver's license, medical) by the application deadline
- Intentionally supplying inaccurate or false information about yourself

---

Job Offer

If you are fortunate enough to receive an offer and you decide to accept it, you have a verbal agreement with the fire department. It will remain a verbal agreement until a written agreement is drawn up and signed.

Congratulations! You are now officially a probationary firefighter!

IF YOU ARE NOT SUCCESSFUL THE FIRST TIME

The process used to select applicants is a rigorous one. Don't be surprised if you don't succeed the first time. It is natural to feel disappointed either with yourself or the process, but you can't let it discourage you. Anything worth doing doesn't come easy. There are so many steps to take, and it is common to be found lacking in some specific area. It is now your job to learn from your mistakes. Here are some tips to help improve your chances of success for next time.

Written Aptitude Test

Applicants can usually re-test after a waiting period. During the waiting period, try to determine the areas that were most challenging for you and prepare yourself to do better on the next test. Some fire departments use categories in the aptitude tests and will, if you ask, identify your weaker areas. These areas that require extra attention can be improved by practice. Books are available which focus on the written aptitude test. If further help is require find a tutor to work with or post your questions in the forum at www.becomingafirefighter.com.

Physical Fitness Test

If you were unable to complete a component of the fitness test, be sure to incorporate exercises in your workout that will help you successfully pass that component the next time. You may want to consult a personal trainer.

Interview

Find out where you went wrong. Most fire departments will be glad to help you get better. What do you have to lose? If you were nervous and you feel you didn't answer the questions as well as you could have, practice in mock interviews with friends, family, or a friendly firefighter. Perhaps you were unprepared and thought you could just "wing it." Next time, make sure you have prepared answers to questions they asked during the first recruitment interview and to others you think they may ask.

Medical Exam

Again, ask the recruiters why you were unsuccessful, and investigate what your options are. You may have to consult your doctor or another medical practitioner.

Unfair Treatment

You may believe you were treated unfairly for some reason, such as racial or gender discrimination, or a harsh judgment on a background check. Find out what the fire department's appeal procedures are and decide whether you want to raise the issue. However, I would make sure you have very strong evidence to support your claim; otherwise, it's not worth the time or money.

Final Word

Write the department, thanking them for the opportunity of applying and for being considered. Tell them you will not give up and that you plan to address the reasons that you were unsuccessful this time around.

KEEPING TRACK OF RECRUITMENTS

Keep track of your recruitment efforts with the Recruitment Manager in figure 11.1. Also keep track of the recruitment pattern and plans of the fire departments you are interested in with the Recruitment Tracker in figure 11.2.

FIGURE 11.1

RECRUITMENT MANAGER

Fire Department	Application Return Deadline	Date Completed Application Package Submitted	Comments

FIGURE 11.2

FIRE DEPARTMENT RECRUITMENT TRACKER

Fire Department	Last Recruitment Date	Projected Recruitment Date	Actual Recruitment Date	Comments

CHAPTER TWELVE

APPLICATION PACKAGES

Once a fire department announces it will be recruiting probationary firefighters, your first step is to pick up an application package. Sometimes, the announcement doesn't state from where or when to obtain them. Your first step is to contact the fire department or the municipality's human resources department to obtain the application. If you haven't actually seen the announcement, make sure you find and read a copy of it. Be sure to check the fire department's website because there may be important information about the recruitment process you may need to know. For example, the Calgary fire department may only be accepting application packages from candidates who reside in Calgary. Other key information you want to be clear about includes dates, deadlines, qualification requirements, whether application forms can be mailed out or picked up, and if there are any fees to be paid. If you're not sure about something, take the time to call and get clarification from either the fire department or the city human resources department.

Fire departments use application forms as a way to standardize the assessment of all applicants' qualifications. Make sure you complete the application in a neat manner and that the information is accurate. Prepare a rough draft on a photocopy, and then transfer your answers onto the original application when you are sure you know what you want to write.

A city's recruitment process can be quite extensive. There can be hundreds of applications to review, and each city sets a minimum standard and creates a marking system that helps in narrowing the applicants down to a manageable number. This could be good news or bad news for you, depending on the depth of your training and experience.

It's important when assembling your application package to put your documents in the proper order. Recruiters don't have the time to search for the information they want.

Unless otherwise stated, the order of your application package should be assembled as follows:

1. Fire department application form (completed and signed)
2. Cover letter
3. Résumé
4. Prerequisite certificates, licenses, and diplomas (Note: Provide color photocopies of course certificates, licenses, and diplomas.)
5. Other, fire-related certificates, diplomas, degrees, and transcripts
6. Other, non-fire-related certificates, diplomas, degrees, and transcripts
7. Receipts for incomplete courses (optional)
8. Copies of previous tests results (optional)
9. Military service record and decorations
10. Awards (academic, work-related, personal achievements, etc.)
11. Criminal record search documents
12. Drivers license and abstract
13. Volunteer documents (location, hours served, responsibilities, etc.)
14. Reference letters
15. Method of payment (cash, certified cheque, money order, etc.)

FIGURE 12.1
APPLICATION PACKAGE CHECKLIST

APPLICATION PACKAGE INCLUDES:

	YES	NO
Application form	☐	☐
Cover letter and updated résumé as of _____ (date)	☐	☐
Prerequisite certificates, licenses, and diplomas	☐	☐
Fire-related certificates, diplomas, degrees, and transcripts	☐	☐
Non-fire-related certificates, diplomas, degrees, and transcripts	☐	☐
Receipts for incomplete courses (optional)	☐	☐
Copies of previous test results (optional)	☐	☐
Military service record and decorations	☐	☐
Awards—academic, work related, personal achievements, etc.	☐	☐
Criminal record search documents	☐	☐
Drivers license and abstract	☐	☐
Volunteer documents (location, hours served, responsibilities, etc.)	☐	☐
Reference letter(s) (only if requested)	☐	☐
Method of payment (cash, certified cheque, money order, etc.)	☐	☐

QUESTIONS TO ASK YOURSELF

	YES	NO
Is my name clearly marked?	☐	☐
Is my contact information clear?	☐	☐
Is all my information accurate?	☐	☐
Are any updates needed?	☐	☐
Are all certificates color photocopied?	☐	☐
Is my cover letter personally addressed?	☐	☐
Is my application package easy to read?	☐	☐
Are the documents and papers in the proper order?	☐	☐
Are the documents and papers in good condition?	☐	☐
Should I submit my application in person or by registered mail?	☐	☐

CHAPTER THIRTEEN

HOW TO SUCCESSFULLY COMPLETE AN APPLICATION

It may seem redundant filling out an application form because you'll find that you are literally copying the information in your résumé onto another piece of paper. But the advantage of an application form to a fire department is that recruiters can choose the order in which they wish to review your information, and the common application form allows them to easily compare applicant's qualifications, since the layout is the same for each applicant. The recruiters can easily pick out the applicants who will advance through the process by quickly glancing at the application form and seeing if all requested information is provided. You could have a very impressive résumé with all kinds of courses, but if you have a lot of empty sections on your application form, then you are more likely to be passed over.

Fire departments that require the application form to be delivered in person usually accept delivery by any person, not necessarily the applicant.

APPLICATION FORM TIPS

Follow the advice below and you will find the application process less challenging.

- Quickly review the entire application form before attempting to fill it out (do not make a mark on it). You will become familiar with the content and structure of the application form, which will help you determine what information goes where.

- Always photocopy your application form before you attempt to fill it out. Use this copy as a work sheet and keep the original as the final, finished form. This way, if you make a mistake, you haven't made a mess of the copy you will be submitting.

- Study and follow all instructions very carefully. Application forms vary from fire department to fire department—and all have specific places where you're expected to fill in your information.

- Complete the application form as neatly as possible. Take your time when filling it out and start early! Don't wait until the night before it is due. If your handwriting is hard to read, then your application may be tossed in the garbage. Remember, your application is a reflection of you.

- Always use either a black or blue pen to fill out your application form. If you make a mistake, whiteout is acceptable.

- Never fold or bend your application form. Keep it in mint condition in a large envelope with all your documentation. If you need to mail your application, use a waterproof and rigid envelope.

- Customize all responses to be relevant to the position of a probationary firefighter. Outline skills and descriptions of your experiences and achievements without including irrelevant or unimportant information.

- Don't leave any box blank. Fire departments structure the application form so that they get the same type of information from every applicant. If you come across a section on your application form that is irrelevant to you, then respond with "not applicable" or "n/a." Never write "see résumé."

- Only include positive information. Never include information that gives them a reason not to give you an interview. If you provide negative information such as you were fired from your last job, you're not sending the fire department the right message. You can address negative information that is raised in an interview, if necessary.

- Honesty is the best policy. Being caught supplying false or misleading information will terminate any chance of getting an interview.

- Your application must be consistent with your résumé. Pay close attention to detail. You want to make sure that all dates, names, course titles, and so on that appear on your application form correspond with the information on your résumé.

- Always verify the information you supply in your application. Don't just estimate the length of time or which years you worked for an employer. Take the time to look up the information so you can be certain it is accurate.

- Thoroughly review your application before submitting it. After you have completed your application form, let it sit for a day or two and come back to it. Chances are you'll notice mistakes such as typos or wording that could be improved.

On the next page, you will find an example of an application form. Keep in mind that actual application forms are larger, providing more space to fill out information.

FIGURE 13.1 • FIREFIGHTER APPLICATION

FIREFIGHTER APPLICATION

PERSONAL INFORMATION

Name:	
Address:	

City:	Postal Code:

Height (without shoes):	Weight:	Do you smoke? (✔) Y ☐ Amount? ___ N ☐

Visual Aids Y ☐ Required? (✔) N ☐	If Yes, describe (i.e. contacts/glasses):
	Visual acuity without aids (i.e. 20/20. Please specify):

Any medical conditions or disabilities that would impair your ability to perform all aspects of the job applied for: Yes ☐ No ☐

If yes, please describe your limitations:

General Health:	Social Insurance Number:		
Driver's License No.:	License Class	Air Brake (✔)	Y ☐
# of Points:	Expiry Date:	Endorsement	N ☐

Are you a Canadian citizen? (✔) If no, are you a landed immigrant? (✔)

Y ☐ N ☐   Y ☐ N ☐

EMPLOYMENT INFORMATION (list most current job first)

1. Company Name:	Type of Business:	
Address:		
Supervisor's name:	Phone #:	Dates employed:
Job Title:		Rate of pay:
Description of duties:		
Reason for leaving:		
2. Company Name:	Type of Business:	
Address:		
Supervisor's name:	Phone #:	Dates employed:
Job Title:		Rate of pay:
Description of duties:		
Reason for leaving:		
3. Company Name:	Type of Business:	
Address:		
Supervisor's name:	Phone #:	Dates employed:
Job Title:		Rate of pay:
Description of duties:		
Reason for leaving:		

EDUCATION AND TRAINING

	School (name/location)	Years Attended (dates)	Highest Grade Completed	Major Subjects	Did you graduate? Y/N
High School					
University or College					
Vocational or Trade Business School					
First Aid Training:					
Expiry Date:					
Can you swim? (✔) Y ☐ N ☐	Certification (provide details/dates)				
Other related training (e.g. scuba certification)					

LEISURE AND RECREATIONAL ACTIVITIES
(specify if past or present activity and time spent on fitness/current interests)

Current leisure/physical fitness activities (describe your fitness program)

Team sports and positions played/playing:

Hobbies/Volunteer work:

REFERENCES
(give three references [not relatives] who have knowledge of your character and ability)

	Name	Address	Phone Number	Occupation	Years
1					
2					
3					

GENERAL

1.	What are your reasons for wanting to become a Firefighter?
2.	Why do you feel particularly suited to this occupation?
3.	Have you made application previously to any fire department (✔) Y ☐ N ☐ If yes, provide details and dates:
4.	Additional information important to your application:

I hereby certify that all information given in this application is true and complete. I agree that any misstatement or misrepresentations herein may cause forfeiture on my part to all rights to any employment with The District of North Vancouver. I further release to the District the authority to check records and files relevant to my applications and to contact employers with regard to work references.

Signature _____   Date _____

ADVANTAGES OF MULTIPLE APPLICATIONS

As discussed previously, applicants must be realistic. You may not be recruited by the fire department of your choice. In fact, it is more likely that you will get hired by the fire department of your third or fourth choice. But once hired somewhere, you can then think about applying elsewhere. Firefighters may stay at their initial job for one or two years before trying to jump ship to another department, perhaps closer to home. So apply to as many fire departments as possible, regardless of where they are located geographically. The more practice you get filling out application forms and writing the aptitude tests, the better you will become at completing them. You may spend a lot of time and money, but ultimately, it's worth it. When the fire department of your choice announces a recruitment, you will be much better prepared to complete the application form and write the aptitude test.

The great thing in this world is not so much where we are,
but in what direction we are moving.
— Oliver Wendell Holmes

CHAPTER FOURTEEN

ORGANIZATION FOR THE RECRUITIMENT PROCESS

Being organized with your firefighting recruitment data is very important for a number of reasons. It enables you to remain focused on what you are doing or what you need to be doing, and it helps you avoid missing application deadlines, and the date and time of your interview. Can you imagine how you would feel if you missed an interview! Reduce stress by focusing on organization.

To remain organized, you will need some type of system. Have a place in your home for everything having to do with "fire." Organize your fire recruitment data so that you will be able to locate information easily. The best way to do this is by using a binder system, one for each topic. For example, use a red binder for fire departments you are interested in. Divide this binder into sections with dividers. For each section, write down the fire department's name, and from then on any information you receive regarding that fire department will go in its appropriate section. You will be surprised how much information you accumulate. When choosing fire departments for your binder, only include the ones you are actively involved in or interested in pursuing. It is unrealistic to list and accumulate information for too many departments, so be selective. Use other binders for other information, such as courses to take and volunteer opportunities.

You may find it useful to make a copy of Figure 14.1 for each department you wish to collect information for. This will help avoid mailing the same fire department twice or, even worse, forgetting to mail them at all. Remember to update this information regularly. Visiting each fire department's website on the same day and time each week will ensure that you stay informed about any changes and any new announcements and other information.

FIGURE 14.1 • FIRE DEPARTMENT PROFILE

Fire department: _____

Phone: _____

Fax: _____

Website: _____

Address: _____

Chief's name: _____

Deputy chief's name: _____

Other contacts (fire halls, etc.) _____

_____

City human resources data: _____

Contact: _____

Phone: _____

Fax: _____

City website: _____

Address: _____

_____

Division of fire districts (jurisdiction): _____

Number of fire halls: _____

Location(s) of fire halls: _____

Population: _____

_____

Type of shifts worked: _____

Number of emergency calls annually: _____

Specialized Teams: _____

Anticipated future growth
    (new fire hall, new trucks etc,): _____

Charitable work undertaken: _____

FIREFIGHTERS MAKE DRAMATIC ROOF RESCUE OF WOMAN TRAPPED BY FIRE

MARCH 19, 2006

Firefighters made a dramatic rescue from the top floor of a burning building in Crown Heights Wednesday morning in a scene that looked like it was straight out of a movie.

Fire and heavy smoke spread to the upper floors of the Ebbets Field Houses in Crown Heights after flames were sparked in a trash compactor in the basement.

The fire forced one woman onto the ledge from the top floor of her apartment. When firefighters arrived at the scene, they found Cheryl Ann John dangling from the roof of the 25-story building, held only by the weakening grip of her husband and a Good Samaritan neighbor.

Three firefighters from Ladder Company 113 and Ladder 132—Joe Donatelli, Bill Hansen and Tim Rail—were able to rappel down from the roof to pull Johnson to safety.

"She ended up letting go of one hand and she put it around my neck, and I don't know if her other hand actually made it around me, but she held on to me, I held on to her, then they pulled me up," said Firefighter Donatelli of Ladder 132.

"This is the life belt that Billy put on," said Firefighter Rail of Ladder 113, holding up a safety belt that made the rescue possible. "He was the anchor. It ties around here [showing metal clip]. This goes over the parapet [holding up belt]. Joe put it on. This is the rest of the rope."

Two people, including the woman who was rescued, were taken to Kings County Hospital with minor injuries.

Three firefighters suffered minor injuries battling the blaze and three civilians, including John and her husband, and another person suffered smoke inhalation.

By: On NY1 Now

When you become aware of a fire department recruitment, you may find it useful to complete the recruitment information form in Figure 14.2 below.

FIGURE 14.2 • FIRE DEPARTMENT HIRING CHECKLIST

Fire department: _____

Phone No: _____

_____

Website: _____

Address: _____

_____

_____

Possible hire date: _____

_____

Hand-in application by: _____

_____

Cost: _____

_____

Prerequisites: _____   _____

_____   _____

_____   _____

You should also have an organized system for keeping track of schools, courses, and licenses. The amount of information can be overwhelming, and a good system for keeping track of options and accomplishments will provide focus and confidence that you are on top of things.

Fire School

If you're planning to attend a post-secondary fire college, use a binder to keep all the information you receive from the colleges you have inquired about. You will also want to make note of related information such as transportation options, accommodation, and financing.

Courses and Licenses

Another binder should be reserved for research you have done regarding courses and training options. Keep a separate section for courses you have completed and licenses you have obtained. By keeping proper documentation of each completed course or license obtained (course description and certificates), you will be clear about the qualifications you have acquired. It may be that you took some of the courses a few years back and you may not remember well the topics covered without reviewing the course outline. Is the material still current, or do you need to update?

Résumé Packages

Another binder is necessary for résumé packages. You should always have an updated résumé on hand. You never know when you're going to need one. You should also keep photocopies of your course certificates and licenses, preferably in color, on hand. Keep copies of your high school diploma, obtain reference letters, and create precedent cover letters and thank-you letters as well.

If you want to take this one step farther, have an up-to-date, pre-assembled résumé package on hand. When a fire department posts a hiring, you will then be prepared. All you will have to do is insert the department's name into the precedent cover letter and on the envelope sticker and you're all set. However, make sure you read the recruitment material carefully and check that your material addresses all requirements.

If what you're working for really matters, you'll give it all you've got.
—Nido Qubein

❗ KEY POINTS TO REMEMBER

- You never know when a fire department will announce a recruitment. Be prepared.
- If you are not successful at any stage of the recruitment process, you can improve your chances for next time.
- Always photocopy the fire department's application form and use it as a draft before you complete the application.
- Be sure to include all necessary documents with your application package.

NOTES

_____

_____

_____

_____

_____

_____

_____

_____

_____

_____

NOTES

NOTES

Part IV

Preparing
for
Written &
Physical
Tests

CHAPTER FIFTEEN

WRITTEN APTITUDE TEST

The firefighter tests are designed to help fire departments assess the aptitude of each applicant. It is specific to entry-level firefighter applicants with no previous fire-related education or experience. It measures applicants' abilities in a number of areas, including their understanding of written and oral information and mathematics, as well as assessing their mechanical aptitude and interpersonal skills.

Usually a fire department will choose one or two days for aptitude testing. It is not uncommon for a fire department to use a large assembly hall as there may be hundreds of applicants writing at the same time. On the day of the test, be sure to arrive early. Before you write the test, you have to register. Fire departments will have a list of candidates who are eligible to write the tests. Often, the candidates will have to provide some photo identification as well as pay a fee. Once registered, you proceed to the room, take a seat, and wait for the test to begin. Instructions will be provided on how to properly fill in the information, such as your name, the test number, the date and time, and the location. You then begin the test.

Scores on the tests are intended to indicate level of ability in areas or skills considered important in the fire service. The scores are helpful in predicting who will be suitable for the profession. The tests are standardized for each fire department's recruitment process, and

Unsuccessful applicants will be returned to the applicants pool for further consideration if other positions become available.

test scores allow comparison of candidates with different experiences and education. Categories commonly found on these tests include any number of the following:

- Memory of oral information
- Mapping
- Understanding written firefighter material
- Grammar
- Mathematical reasoning
- Understanding three-dimensional diagrams
- Construction
- Tool recognition
- Teamwork
- Community living

The test you write could cover all of these areas or only a few. It depends on what areas are important to the fire department you are applying to.

The CPS (Cooperative Personal Services) test is the most common entry firefighting test you will find in Ontario. The rest of Canada and the United States use tests that are similar to the CPS test. It consists of 100 questions and is made up of five different categories:

- Oral comprehension—a passage will be read to the group that is less than 10 minutes in length, or there may be two smaller passages read back to back, each ranging from 1 1/2–3 minutes (20 questions)
- Reading comprehension (25 questions)
- Mathematics (20 questions)
- Mechanical aptitude (20 questions)
- Interpersonal skills (15 questions)

Candidates must pass the test in order to advance and be considered for an interview. In order to pass, candidates must attain a score equal to or higher than a set threshold score, which is usually 70 percent or higher. In order to give applicants a breakdown of their marks, some fire departments will score the categories separately. The sum of the scores in each category equals the total score. Notice in the scored example below how reading comprehension is weighted the heaviest.

- Oral passage   16/20
- Reading comprehension   18/25
- Mathematical   16/20
- Mechanical aptitude   14/20
- Interpersonal skills   14/15

TOTAL SCORE   78/100

Applicants must attain a set threshold score in the aptitude test.

Most, if not all, fire departments use a written aptitude test, and for those candidates applying to several fire departments, you will find yourself writing several tests. Some fire departments may use the same entry test, but for the most part, the tests will be different. Most fire departments now contract the responsibility of creating entry firefighter tests to independent businesses that create tests for employers. However, some fire departments continue to create their own tests and have had great recruitment success in doing so.

You don't have to be an experienced firefighter to do well on these tests. For example, with a grade 12 education or equivalent, you should be able to pass the mathematical portion of the test. However, it is advisable to prepare for these tests, and Chapter 16 addresses how to accomplish this.

CHAPTER SIXTEEN

PREPARING FOR AND WRITING THE APTITUDE TEST

To help relieve test anxiety and to improve your chances of choosing the correct answers, I recommend writing as many practice entry-level firefighter tests as possible. How would you feel if the fire department you've been waiting for posts a recruitment and you've taken many courses and done lots of community work, but you write the aptitude test and fail!

PREPARING FOR THE TEST

Here are some key points when preparing for your test:

- Highlight or circle the areas in the questions you have problems with. Continue to go over these areas until you understand what is being asked.

- It is important to free yourself from distractions, such as music, television, or people talking near you.

- Adequate lighting helps increase your concentration when studying.

- If you smoke, you might find it difficult to last two to three hours without having a cigarette. You should practice not smoking for this period of time, so you won't be anxious about not having a smoke during the examination. (You should quit!)

- Exercise not only helps the body, but also the brain. Be sure to stay in shape, and try to avoid stressful situations to keep your mind clear at examination time.

- Rest is extremely important, not only the night before the test, but the week before as well. Try to be consistent with your sleeping habits. For example, if you are accustomed to getting six hours of sleep a night, be sure to get six hours or more the night before the test. But don't overdo it. Too much sleep the night before may leave you feeling groggy.

- Make sure that on the day of your test, you leave early and allow extra time for traffic problems. Be in the examination room 20 to 30 minutes ahead of time. Allow yourself the time to become accustomed to the temperature of the room. If you're too hot or too cold, you can address the problem in time to write the test. Layer your clothing to make sure you are comfortable.

- The examiner will hand out the tests and give specific instructions, such as how to complete your answers, what the time limits are, and how to respond to the timing signals. If at any time you do not understand any of these instructions, raise your hand to ask a question. Don't cut in when someone from the department is speaking. Make sure you understand exactly what to do, and how long you will have to do it. Stay calm.

WRITING TIPS

Listening Skills

Sit near the front of the room to eliminate distractions and help you focus and listen. The listening portion of the test proves to be the most difficult for many candidates. It's usually the first part of the test, and chances are you are nervous. Also, you may find yourself concentrating on the specific details in the passage, and you may miss the whole picture. There are things you can do to improve your comprehension.

- It's easier to remain focused during the oral reading of the passage if you maintain eye contact with the reader. This will help to avoid distractions and increase your concentration while the passage is being read aloud. You can't afford to miss any information.

- Concentrate on the content, not the delivery. It's easy to be distracted if you find the reader amusing because of the way he or she speaks. Ignore such things as the hand gestures used or the number of times the reader blinks during the passage. Focus only on the content.

- Don't get emotionally involved in the passage; you may make the mistake of hearing what you want to hear and not what is actually being said. Remember not to be biased or opinionated. Keep an open mind.

- Continue to control your body temperature by removing or adding layers of clothing during the test.

- Control the time between your rate of thought and the reader's rate of speech. Did you ever wonder why you lose your concentration so easily? It's because your mind can think faster than someone can speak. To avoid becoming bored, try to anticipate the end of the sentence. It's possible to listen, think, and write all at the same time. It just takes practice.

When answering a question, eliminate choices that:

- benefit you, the firefighter, alone;
- call for unnecessary actions that are contradictory to what is required;
- only partially solve the problem; and
- insult, disregard, interfere with, or place in peril a citizen

Reading Comprehension

The reading comprehension section of the test measures both literacy and your understanding of the text. You will be provided with a short reading passage followed by several multiple choice questions to test your comprehension. All the answers to the questions can be found in or deduced from the passage itself. The passages usually consist of technical information—nothing like a novel or essay. You may find yourself pressed for time during this portion. To help you concentrate and read faster, use your pencil to follow the words as you read along. This helps you focus on the details and increases your concentration on the task at hand.

Take the time to circle or underline the key words and phrases that you feel are important in the passage. This will allow you to focus on the important information required to answer the questions. For short passages that are only six to seven sentences in length, take the time to read the passage carefully, so that you only have to read it once. You should be able to retain all of the essential information in the first reading.

Before reading longer passages, take a moment to peruse the questions you are being asked. This helps you to concentrate on the relevant parts of the passage. Also, your memory works better if you can visualize what you read—or, better yet, place yourself in the passage. By placing yourself in the passage, you will relate more to the information and find it easier to remember.

For longer passages, you may find it easier or quicker not to read the passage once through first, but to just look for the information asked in each question. Everyone thinks and works differently. To see what works best for you, you can test various approaches by having a friend find a newspaper article and write questions about the content. Then, the person can give you the article and have you answer the questions. The more you practice answering questions on reading passages, the more confident you will be when taking the test.

What we see depends mainly on what we look for.
—Sir John Lubbock

Scenario:

Fire extinguishers are probably the most common fire-protection appliance used today. You can find fire extinguishers in most homes, commercial businesses, industrial work places, and commercial vehicles. Types of fire extinguishers may vary but they all share the same principal, fire extinguishment. Fire extinguishers are only effective for extinguishing incipient fires. Often a fire extinguisher can extinguish a small fire in less time than it would take to deploy a hose line to put out the fire. There are regulations, such as NFPA 10 (standard for portable fire extinguishers) that regulate the rating, selection, and inspection of portable fire extinguishers.

There are different types of portable fire extinguishers: some common ones are pump tank water, stored pressure water, halon, carbon dioxide, dry chemical, and dry powder. All are effective if used on the proper classification of fire. Without selecting the proper fire extinguisher, you may cause more damage than good.

There are four classifications of fire. A class A fire is one with ordinary combustibles, such as paper and wood. A class B fire is flammable liquids such as gasoline or oil. The third is a class C fire, energized electrical fires (if power is de-energized, the class C fire turns into a class A fire). The fourth classification of fire is a class D fire, which involves combustible metals such as magnesium and / or sodium. Portable fire extinguishers that are suitable for more than one class of fire are identified using different combinations of letters A, B, and C. The most common types of combinations are class ABC, class AB, and class BC. Portable fire extinguishers must be properly marked or the extinguisher should not be used.

Along with the class of fire that the extinguisher will be used on, it should also have the amount of extinguishing agent. For example, the fire extinguisher may have a rating of 4-A 20 BC. The extinguisher should extinguish a class A fire and extinguish 4 times larger than a 1-A fire, extinguish 20 times as much class B fire as a 1-B extinguisher, and extinguish a deep layer of a 20-square-foot (2-square-meter) area. This extinguisher can also be used on fires involving energized electrical equipment.

These fire extinguisher ratings are for untrained users. A trained individual would be able to put out a greater amount of fire. To properly select a portable fire extinguisher, there are things you need to consider, such as classification of the burning fuel, rating of the extinguisher, hazards to be protected against, severity of the fire, atmospheric conditions, availability of trained personnel, ease of handling extinguisher, and any possible life hazards or operation concerns.

When using an extinguisher, remember the acronym PASS: P–pull the pin, A–aim, S–squeeze the trigger, and S–sweep back and forth over the fire. Using an extinguisher can prevent a little accidental grease fire in a frying pan from becoming one that burns a house to the ground. Make sure that if you have an extinguisher, you know its limitations and become familiar with how to use it.

1. Where else can portable fire extinguishers be found besides homes, commercial businesses, and industrial work places?
   A) domestic vehicles
   B) commercial vehicles
   C) motor homes
   D) boats

2. Fire extinguishers are effective in extinguishing what types of fires?
   A) garbage fires
   B) fully developed fires
   C) Incipient fires
   D) free-burning fires

3. Fire extinguishers can put out small fires
   A) in the same time it takes to deploy a hose line.
   B) in less time than it takes to deploy a hose line.
   C) in more time than it takes to deploy a hose line.
   D) only if the fire is outdoors.

4. The regulation for portable fire extinguishers is
   A) NFPA 12.
   B) NFPA 10.
   C) NFPA 15.
   D) NFPA 20.

5. What are the four classifications of fire?
   A) 1, 2, 3, and 4
   B) Orange, yellow, black, and green
   C) A, B, C, and D
   D) None of the above

6. What are class A fires?
   A) ordinary combustibles
   B) energized electrical
   C) flammable liquids
   D) combustible metals

7. What are class C fires?
   A) ordinary combustibles
   B) energized electrical
   C) flammable liquids
   D) combustible metals

8. What are the three combinations of fire extinguishers?
   A) class A, class BC, class AB
   B) class ABC, class C, class AB
   C) class ABC, class AC, class BC
   D) class ABC, class AB, class BC

9. What do you need to consider, when selecting a fire extinguisher?
   A) classification of burning fuel
   B) severity of the fire
   C) both A and B
   D) none of the above

10. What does the acronym PASS stand for?
    A) pull pin, aim, safety glasses, and stance
    B) pull pin, aim, squeeze trigger, and sweep
    C) pick up extinguisher, aim, squeeze, and sweep
    D) pull pin, aim, squeeze, and safety

Answers:
1. B; 2. C; 3. B; 4. B; 5. C; 6. A; 7. B; 8. D; 9. C; 10. B

CHAPTER SEVENTEEN

STRATEGIES FOR ANSWERING MULTIPLE CHOICE QUESTIONS

When you take the test, you will be required to mark your answers on a sep-arate machine-scored answer sheet. Total testing time is usually two hours in length; there are no sections that are separately timed. The following are some general strategies you may want to consider.

- Look over the entire test before attempting to write it. Make sure you are not missing any pages and that you are writing the right test.

- Read the test directions carefully, and work as rapidly as you can without being careless. For each question, choose the best answer from the available options.

- Questions for which you give no answer or give more than one answer are counted as incorrect answers.

- If you have no idea what the correct answer is, but you lose the same amount of points for blank answers as incorrect answers, then you may as well guess.

- If you decide to change an answer, make sure you completely erase it and fill in the spot corresponding to your desired answer. Most tests are graded by computer and the computers aren't designed to figure out which of two answers you meant.

- Never wait until the last few minutes to record your answers. You may become distracted and not get your answers recorded on time.

- Eliminate options you know to be incorrect. If allowed to mark the test paper, cross out options you know to be incorrect. Apply the "true-false test" to options you are unsure of.

- You may want to work through the test more than once, first answering questions that are easy or about which you feel confident. Then, start again to answer questions that require more thought, then again, finishing with the most difficult questions until the time runs out.

- If you think two of three options are correct, "all of the above" is likely the correct answer.

- If you are unsure about answers with numbers, eliminate the high and low numbers and consider the middle-range numbers.

- Double negatives are difficult to decipher. Reword it into a positive statement and reconsider.

- If two options are opposite each other, chances are one of them is correct.

- Favor options that contain qualifiers. Answers that include more information are often correct.

- Use hints from other questions to help you answer the question.

Selecting the correct answers on a multiple choice test can be difficult at times. Usually you can eliminate two of the four answers right away. The tricky part is now deciding which one out of the two answers remaining is the correct one. Paying close attention to the wording of the answers can give you hints about whether an answer is more than likely correct or incorrect. If the answer limits or overstates, it is probably incorrect. If the answer is less exact or restrictive, it is probably correct.

Example:

> A firefighter is required to wear a PASS (personal alarm safety system) alarm at any fire scene where the firefighter is required to enter any structure wearing SCBA. If a firefighter's PASS alarm fails to work properly, the officer in charge will be notified of the problem immediately. Chances are the firefighter will be reassigned to a low-risk task such as managing the accountability board until the situation has been brought under control or the firefighter gets the PASS alarm working properly and is reassigned back to the structure.

Based on the paragraph above, select the correct answer.

A) A firefighter should wear the PASS alarm only when entering a structure wearing SCBA.

B) A firefighter must attempt to repair a PASS alarm which is not working properly.

C) A firefighter whose PASS alarm is not working properly will be assigned to the accountability board.

D) The firefighter whose PASS alarm has been repaired can be reassigned back to the structure.

Certain words in the above answers can help determine whether the answer is correct or incorrect.

- A) is incorrect because it contains the word "only." Note: Without the word "only," A) would be the correct answer.
- B) would be correct if it said "may" instead of "must."
- C) is incorrect because it says "will." Note: C) would be correct if it said "may" instead of "will."
- D) is correct; it uses the word "can." Note: D) would be incorrect if it said "must" instead of "can."

When selecting your answer, be sure to watch for words such as the following that limit or overstate. These words may indicate the wrong answer.

- no one
- all
- every
- certainly
- always
- will

- invariably
- surely
- any
- no matter
- nothing
- ever

An answer is more likely to be correct if it includes words that are less exact or restrictive, such as the following.

- many
- sometimes
- may
- some
- possibly
- can
- often

- might
- usually
- could
- occasionally
- generally
- probably

THE LITTLEST FIREFIGHTER

In Phoenix, Arizona, a 26-year-old mother stared down at her 6-year-old son, who was dying of terminal leukemia. Although her heart was filled with sadness, she also had a strong feeling of determination. Like any parent, she wanted her son to grow up and fulfill all his dreams; now that was no longer possible. The leukemia would see to that. But she still wanted her son's dreams to come true. She took her son's hand and asked, "Billy, did you ever think about what you wanted to be once you grew up? Did you ever dream and wish what you would do with your life?"

"Mommy, I always wanted to be a fireman when I grew up."

Mom smiled back and said, "Let's see if we can make your wish come true." Later that day she went to her local fire department in Phoenix, Arizona, where she met Fireman Bob, who had a heart as big as Phoenix. She explained her son's final wish and asked if it might be possible to give her six-year-old son a ride around the block on a fire engine.

Fireman Bob said, "Look, we can do better than that. If you'll have your son ready at seven o'clock Wednesday morning, we'll make him an honorary fireman for the whole day. He can come down to the fire station, eat with us, go out on all the fire calls, the whole nine yards! And if you'll give us his sizes, we'll get a real fire uniform for him, with a real fire hat—not a toy one—with the emblem of the Phoenix Fire Department on it, a yellow slicker like we wear and rubber boots. They're all manufactured right here in Phoenix, so we can get them fast."

Three days later Fireman Bob dressed him in his fire uniform and escorted him from his hospital bed to the waiting hook and ladder truck. Billy got to sit on the back of the truck and help steer it back to the fire station. He was in heaven. There were three fire calls in Phoenix that day and Billy got to go out on all three calls. He rode in the different fire engines, the paramedic's van, and even the fire chief's car. He was also videotaped for the local news program. Having his dream come true, with all the love and attention that was lavished upon him, so deeply touched Billy that he lived three months longer than any doctor thought possible.

One night all of his vital signs began to drop dramatically and the head nurse, who believed in the hospice concept that no one should die alone, began to call the family members to the hospital. Then she remembered the day Billy had spent as a fireman, so she called the Fire Chief and asked if it would be possible to send a fireman in uniform to the hospital to be with Billy as he made his transition.

The chief replied, "We can do better than that. We'll be there in five minutes. Will you please do me a favor? When you hear the sirens screaming and see the lights flashing, will you announce over the PA system that there is not a fire? It's just the fire department coming to see one of its finest members one more time. And will you open the window to his room?"

About five minutes later a hook and ladder truck arrived at the hospital and extended its ladder up to Billy's third floor open window; 16 firefighters climbed up the ladder into Billy's room. With his mother's permission, they hugged him and held him and told him how much they loved him.

With his dying breath, Billy looked up at the fire chief and said,

"Chief, am I really a fireman now?"

"Billy, you are, and the Head Chief, Jesus, is holding your hand," the chief said.

With those words, Billy smiled and said, "I know, he's been holding my hand all day, and the angels have been singing..." Then he closed his eyes one last time.

— Julia Meadows, *Landmarks Magazine*

CHAPTER EIGHTEEN

PHYSICAL FITNESS TEST

Firefighting is one of the most physically demanding professions. Firefighters require high levels of physical fitness to perform and execute all fire ground operations and tactics. Firefighters must maintain a healthy body core temperature, blood pressure, and heart rate when exposed to a hot environment for a sustained period of time.

Fire departments use physical fitness tests to ensure firefighter recruits have an above-average fitness level in flexibility, cardiopulmonary endurance, muscular strength, and muscular endurance.

Some fire departments will offer you a job on the condition that you pass a physical fitness test, while other fire departments have you take a physical fitness test prior to offering employment. These tests are usually valid for six months and cost from $120 to $200. Some candidates will take the test every six months to ensure they have it if they need it, but this is an expense that many can't afford. You can certainly wait until you are asked to undergo one. In most recruitment situations, you are given enough notice to make an appointment.

A couple of different types of physical fitness tests are available, and although each testing facility offers similar tests, not all are recognized by each recruiting fire department. It is possible that if you apply to three different fire departments, you may have to take three different physical fitness tests. However, chances are, if you pass one you will pass them all.

It is your responsibility to prepare for your test. Some places that offer the tests provide help to get you into shape. Make inquiries to determine which physical fitness test you will have to take and exactly what tasks you will be asked to perform. Talk to candidates who have already completed the test. Find out which parts will be most challenging for you, and adjust your workout to incorporate specific exercises that will be of benefit. If you aren't sure which exercises to incorporate into your workout, ask a fitness trainer at your local gym.

You see things; and you say "Why?"
But I dream things that never were; and say, "Why not?"
— George Bernard Shaw

CPAT—PHYSICAL FITNESS TEST

CPAT (Candidate Physical Aptitude Test) will eventually become the standard fitness assessment test throughout Canada. CPAT uses a pass / fail method of grading each candidate; this is based on completing each event in a particular order as well as completing all events within the allocated time frame of 10 minutes and 20 seconds. CPAT is unique because each prop has been clinically tested to ensure consistency and safety as well as to provide fire departments with an accurate assessment of simulated firefighting tasks. Throughout the duration of the test, each candidate must wear additional weights to simulate bunker gear (50-pound or 22.8-kilogram vest) and firefighting equipment (self-contained breathing apparatus). To simulate a high-rise pack, the stair climb is the only event in which an additional 25 pounds (11.3 kilograms) is used.

Each candidate participating in the fitness test will be assigned an evaluator to provide guidance, advise about time, and evaluate the performance of each event. Candidates mimic in order a sequence of actual fire ground activities. Between each event, candidates are assigned an 85-foot (26-meter) walk. This time (20 seconds) is used to recover and prepare for the next event.

CPAT consists of eight events in the following order:

1. Stair climb
2. Hose drag
3. Equipment carry
4. Ladder raise and extension
5. Forcible entry
6. Search
7. Rescue
8. Ceiling breach and pull

1. Stair Climb

This event is designed to simulate climbing four flights of stairs in full protective clothing while carrying a high-rise pack.

- Candidates must carry an additional 25 pounds (11.3 kilograms) (simulated high-rise pack).
- Once the warm-up is completed, the three-minute test of 60 steps per minute will begin. Candidates may only use the hand rails during the test to re-establish balance.
- Then, you walk 85 feet (26 meters) to the next event.

2. Hose Drag

This event simulates the dragging of an uncharged hose line from the fire apparatus to the fire.

- Candidates must drag a 200-foot (61-meter) length of 1 3/4-inch (4.5-centimeter) fire hose.
- Then, you must drape a nozzle attached to 100 feet (30.5 meters) of 1 3/4 inch (4.5-centimeter) hose over your shoulder or across your chest and, dragging in 75 feet (23 meters), make a 90-degree turn and continue dragging the hose an additional 25 feet (8 meters) to a mark indicated on the ground.
- Candidates must then stop and pull 50 feet (15 meters) of hose toward themselves from a kneeling position past the mark on the ground.
- Then, you walk 85 feet (26 meters) to the next event.

3. Equipment Carry

This event simulates retrieving power tools from the fire apparatus, carrying them to the emergency scene, and returning them to the fire apparatus.

- Candidates remove two power saws from a cabinet, one at a time, and set them on the ground.
- Then, you pick up the saws and carry them 75 feet (23 meters) to a marker, turn the marker around, and then carry them back to the cabinet.
- Then, you place the tools back in the cabinet.
- Then, you walk 85 feet (26 meters) to the next event.

4. Ladder Raise and Extension

This event simulates the placing of a ground ladder at a structure fire and extending the ladder to the roof or a window.

- The event involves two 24-foot (7-meter) extension ladders.
- Candidates must first raise the extension ladder by placing the butt of the ladder against the wall and lifting the ladder over their head at the opposite end. Then, they walk toward the wall using a hand-over-hand method, gradually raising the ladder until it is upright (vertical).
- Once the first ladder is raised, candidates must quickly take hold of the second ladder and extend the fly of the ladder to reach a predetermined height and then bed (or lower) the ladder back to its original position.
- Then, you walk 85 feet (26 meters) to the next event.

5. Forcible Entry

This event simulates the forcing of a locked door or the breaching of a wall.

- Candidates use a 10-pound (4.5 kilogram) sledge hammer to strike the side of a device mounted 39 inches (1 meter) off the ground.
- A buzzer indicates when to stop.
- Then, the sledge hammer is placed on the ground.
- Then, you walk 85 feet (26 meters) to the next event.

6. Search

This event simulates searching for a fire victim with limited visibility in an unpredictable area.

- Candidates start off crawling through a dark tunnel 3 feet (0.9 meters) high by 4 feet (1.2 meters) wide.
- The tunnel is 64 feet (20 meters) in length with two 90-degree turns.
- Candidates must advance through this tunnel navigating over and under obstacles along the way. There are two locations in the tunnel where candidates must crawl through a narrow space.
- Once exiting the tunnel, you walk 85 feet (26 meters) to the next event.

7. Rescue

This event simulates the removal of a victim or injured firefighter from a fire.

- Candidates must drag a weighted mannequin 45 feet (14 meters) to a marker and then make a 180-degree turn around the marker and return 35 feet (11 meters) to the finish line.
- You may release the mannequin to get a better grip.
- To complete the course, the mannequin must be brought completely across the finish line.
- Then, you walk 85 feet (26 meters) to the next event.

8. Ceiling Breach and Pull

This event is designed to simulate the breaching and pulling down of a ceiling to check for fire extension.

- Candidates must remove the pike pole from the bracket.
- Then, you stand under a simulated ceiling and extend the tip of the pike pole to push open a 60-pound (27-kilogram) hinged door three times.
- Then, using the hook, you must pull down an 80-pound (36-kilogram) device from the ceiling five times.
- You must repeat these events four times.

PREPARING FOR THE PHYSICAL FITNESS TEST

Listed below are some simple, effective ways to prepare for your physical fitness test.

- To decrease the chance of cramps during your test, be sure to drink lots of water the day before the test. Drink water until you have to urinate 15 minutes after taking another drink. Make sure you drink at least one liter of water one hour before your physical fitness test starts.
- To increase your level of fitness, you must be in good cardiovascular condition. You want to be able to achieve a heart rate of between 140 and 160 beats per minute. You should exercise to keep this heart rate for 20 to 30 minutes by jogging or other exercise.
- Take the time to warm up before each type of exercise. For the best results, your warm up should simulate the type of exercise. For example, if you're preparing to run, you should run in place for at least two minutes or for a short distance at a very easy pace.
- Stretching for 10 minutes before and after each type of exercise is recommended. Hold each stretch for 10 seconds in a range of motion that produces only mild tension. Then, you should move slightly farther, to the point where you feel a little more tension. Hold this for another 10 seconds.

Test day is not to be taken lightly. Make sure you are there on time—you don't want to miss important instructions, or worse, be disqualified. Have water with you, and take a good source of energy, such as a nutritional bar. Often, shower facilities are available for you to use on site, so bring a towel and a change of clothing.

And don't take the attitude that you are in this alone. It will be a long day with lots of waiting time, so make the effort to be friendly; it's always nice to have someone encouraging you during the test. The camaraderie can even be overwhelming—that's what firefighting is all about. Chances are, there will be some people who are not taking the test that day for the first time. It may be the third or fourth time for some, so if you have any questions, chances are someone with more experience will be able to help you.

When using ankle weights, make sure they are tight around your ankles. If you become fatigued, loose ankle weights will hinder your natural step and could lead to injury.

Be certain that you fully understand every instruction given. Listen very carefully and concentrate! If you are uncertain, don't be shy. It's better to ask questions and show your confidence than to perform the exercise incorrectly. If the equipment you have chosen doesn't fit you right, get some that does; if you fail a task because your equipment failed you, you will regret it. It is important to take control over things that you can. If you are prepared in mind and body, you will have a greater chance for success.

FIREFIGHTER FUND RAISES MONEY FOR NEW ORLEANS COLLEAGUES

Darryl Wiechman, a volunteer fire fighter for the RM of MacDonald (just south of Winnipeg) is selling t-shirts to fire fighters across the country to raise money for New Orleans' beleaguered fire fighters, 80 per cent of whom are still homeless as a result of last fall's Hurricane Katrina. The t-shirts feature a fire truck coming out of a circle on the front and the slogan, "Brothers Helping Brothers" on the back.

Wiechman is selling the t-shirts for $20 each (plus freight) and says that 100 per cent of the money is going to New Orleans' Bravest, a fund set up by New York fire fighters to help their brethren in New Orleans. Wiechman notes that he has almost sold out of his first run of 1,000 t-shirts. To order, readers can contact Wiechman at darrylonfire@mts.net.

Fire Fighting in Canada. February 2006, p. 18

CHAPTER NINETEEN

STAYING MOTIVATED FOR YOUR WORKOUTS

Sometimes the hardest thing about working out is actually making it to the gym. We'd all prefer to be in shape, and certainly firefighting requires it. Very few careers require the strength, stamina, and bravery on such short notice that firefighting does. At the sound of an alarm, whether it's in the middle of the day or at 4:00 a.m., we are expected to be at the top of our game instantly. Being in great physical shape not only helps us keep up with the physical and mental demands of the job, but also helps prevent injuries, both on and off the fire grounds. The last thing you need to happen after you get hired is to end up hurting yourself, thus preventing you from making it through your probation period.

Amateurs practice until they can do it right; pros practice until they can't do it wrong.

Dreams and dedication are a powerful combination.
— William Longwood

Choosing a workout can be challenging and intimidating. There are so many styles and types of workouts, it is difficult to determine which workouts will get you the best results. Most workouts are based on the same principles. Exercise puts physical stress on the body, which boosts metabolism, encourages weight loss, builds lean muscle, provides energy for the day, provides alone-time during the day, relieves stress and body tension, and re-educates muscles that have become weak from injuries or just plain lack of use.

For the best results, you must have a routine that will best suit you in terms of time commitments and your current level of fitness. Advancing too quickly could cause you to become discouraged or cause injury. When working out it's important that you concentrate and enjoy it. Otherwise, you'll start making excuses for why you can't work out to avoid it. My advice is, if you don't like your workout, find another one. It could be very helpful to consult a trainer.

Setting short-term and long-term goals help keep you on track and define what you want to achieve. Here are some things to keep in mind that will help you stay motivated and achieve your goals.

- Make sure your goals are measurable. A goal like "I want to be strong" is too general and hard to measure. But if you say "I want to bench press 275 pounds (125 kilograms) for 6 repetitions by the time of my fitness test on April 6," you are stating a measurable goal.

- Don't set goals that are unattainable—be realistic. You will become frustrated and chances are you will want to give up. On the other hand, if goals you have set are too easy, you won't get the physical results you need, and this will be frustrating as well.

- To achieve your long-term goal, you'll have to set short-term goals to give yourself the necessary stepping stones. For example, if your long-term goal is to squat 300 pounds (136 kilograms), you should set some short-term goals, such as increasing your weight weekly or monthly by five or ten pounds until you reach your long-term goal. Take your time to build muscle slowly and avoid injury.

- Make your workouts enjoyable. If your attitude is to have fun while you work out, you are less likely to fall short of your long-term goal. Try listening to music that is upbeat or allow yourself a favorite treat only after your workout.

- Add variety. You can lose concentration if you become bored with your workout. If there is a particular exercise you don't like, vary how often you do it or the amount of repetitions you do. And there are many different machines or exercises that will produce the same result.

- Find a friend to work out with, but make sure the person takes it as seriously as you do or you will have problems keeping both of you motivated.

- Sometimes people become bored with their workouts. You may feel that you've reached a plateau. This is a normal part of working on long-term goals. You may have to adjust your exercises, sets, and number of repetitions. Or research new workouts—you may find one that better suits you, or that you like as a change.

- Make a strong commitment by scheduling your workout so that they become a fixed daily activity. Try to work out at the same time every day. It will eventually feel unnatural if you don't do your workout at that scheduled time.

- Get serious and be organized. Keep a record. Make a chart to record your achievements every day. Record date, time, repetitions, and weight. You'll soon see your progress and the sense of accomplishment will be most rewarding. You can use a chart similar to the one in Figure 19.1 below.

- Remember—adequate sleep and a good diet is necessary to achieve a good fitness level.

FIGURE 19.1 • WORKOUT LOG

WEEK:										
EXERCISES	**SET 1**		**SET 2**		**SET 3**		**SET 4**		**SET 5**	
DATE	WEIGHT	REPS	WEIGHT	REPS	WEIGHT	REPS	WEIGHT	REPS	WEIGHT	REPS
1.										
2.										
3.										
4.										
5.										
DATE	WEIGHT	REPS	WEIGHT	REPS	WEIGHT	REPS	WEIGHT	REPS	WEIGHT	REPS
1.										
2.										
3.										
4.										
5.										

❗ KEY POINTS TO REMEMBER

- Write as many practice written aptitude tests as you can—practice makes perfect.
- A variety of tests are used. Some fire departments create their own written aptitude tests, but most use external test-writing agencies.
- Read over the entire test before you begin to write it.
- Get lots of sleep the night before the test.
- Be sure you know exactly where your test is taking place—being late would be disastrous.

NOTES

_____

_____

_____

_____

_____

_____

_____

_____

_____

_____

_____

_____

NOTES

NOTES

Part V

Acing the Interiew

THE INTERVIEW

Fire departments use interviews to help determine which applicants have the personality and characteristics they're seeking. Interviews are not intended to verify the skill level of the applicant. Generally, your accomplishments, such as fire experience, education, and work experience are the reasons why you were invited for an interview. The interviewer's main purpose is to assess your interpersonal skills, communication skills, integrity, judgment, decision-making abilities, respect for diversity, and adaptability. From your perspective, interviews give you an opportunity to learn about and assess the fire department and the city you are applying to.

WHAT THEY'RE LOOKING FOR

Fire departments are looking for someone who:

- Has a professional appearance and attitude
- Is passionate about becoming a firefighter
- Is physically fit, healthy, and self-confident
- Exhibits maturity and composure—someone who can handle responsibility, with the life experiences to back it up

Never underestimate the power of professional appearance.

- Has great interpersonal skills and is a team player
- Is modest, sincere, and compassionate
- Has a strong work ethic, hands-on skills, and mechanical aptitude
- Understands the importance of following orders, guidelines, and procedures;
- Demonstrates leadership skills from experience in various projects and organizations
- Has strong communication skills
- can resolve complicated problems in a team environment
- Is able to adapt to new environments and settings
- Is capable of assuming a high level of responsibility at a moment's notice
- Is an upstanding citizen who has respect for the law and all members of the community.

Most interviews are conducted before a panel. The panel members often consist of representatives of the fire department and human resource personnel from the municipality. It is easy to become intimidated in this type of environment, but if you know what to expect, and you prepare well for the interview, you will be less nervous, more confident, and you will make a more favorable impression.

TELEPHONE INTERVIEWS

It is possible that a first interview will take place over the phone: then, if the phone interview is successful, you will be invited for a face-to-face interview. Fire departments have started to use phone interviews as an initial employment screening technique for a variety of reasons. First, they save a lot of time—fire departments are able to screen a greater number of candidates. Second, it is more practical to screen out-of-town and out-of-province or -state applicants this way.

Typically, a phone interview will last 20 to 30 minutes. There could be one interviewer or several via a conference call. You prepare for a phone interview the same way you would for a face-to-face interview. However, there are additional things to consider.

- Treat the phone interview seriously, just as you would a face-to-face interview. It may seem less formal, but it is just as important.

- Have your résumé and cover letter in front of you for quick reference. You can be certain there will be questions regarding its content.

- Be sure to have a note pad and several pens that work. This way you can make notes. Jot down the names of the interviewers, the questions being asked, and the answers you provide. This way, if you are invited for a second interview in person, you can review what you said in the previous interview. This will ensure your answers are consistent.

- Make a cheat sheet. As a point of reference, this will help you respond to questions asked, and prompt you to mention key points you want the interviewers to know about you.

- Use a high-quality phone: Make sure you use a landline—this will prevent your cell phone from cutting out or your cell phone battery failing during the interview.

- Look the part, even though they can't see you. Be sure to clean up and shower prior to the phone interview. This will help you prepare mentally as well as improve your self-confidence. Be sure to be sitting and not lying down on your bed.

You may not consider yourself "good on the phone," and feel dread at the upcoming phone interview. But remember, the other candidates will likely be feeling the same way. Harness that negative energy and use it to prepare yourself—then you can ace it!

CHAPTER TWENTY-ONE

HOW TO PREPARE FOR THE INTERVIEW

RESEARCH

It is vital to take the time before your interview to research general facts about the city and the fire department. Obtaining this information greatly improves your chances of having a successful interview.

You should know the following facts about the city, municipality, or region:

- Mayor's name
- Population (different seasons)
- City attractions—what makes it unique?
- Natural hazards (escarpments, water, etc.)
- Breakdown of zoning (industrial, commercial, residential)
- Plans for expansion or annexation in the future
- Number of hospitals and locations
- Location of airport, train and bus stations, subway, etc.
- Potential man-made hazards (chemical plants, mines, etc.)

You should know the following facts about the fire department:

- Chain of command
- Chief's name
- Deputy chief's name
- Fire districts, number of fire halls, and their location in the city (try to visit each one, which will help to commit locations and neighborhoods to memory)
- Type of shifts worked
- Charities in which it's involved
- Specialized teams (high angle, hazardous materials, water rescue, etc.)
- Number of calls annually (at each hall, if possible type)
- Daily station duties
- Salary, vacation time, and benefits
- Public education involvement (home safety visits)
- Mutual or automatic aid policies (Mutual aid is where neighboring townships agree or have an understanding to respond to each others' fire/emergencies when requested. Automatic aid is the same as above however for specific residences or addresses both townships will respond automatically.)
- Number of firefighters employed by the city
- Department history (department website, city hall archives, public library)
- Is it a composite fire department or integrated EMS (Firemedics). Composite fire departments have both full-time and part-time (volunteer) firefighters
- Anticipated future growth (new fire halls, trucks, etc.)
- Details of any recent major fires or incidents (visit the sites)

TIP!

Contact the city's tourism information center. Have them send you any information they have about the city, including maps, brochures, contact numbers, and information on real estate, shopping, schools, services, festivals, and so on.

SAMPLE INTERVIEW QUESTIONS

Review the following sample interview questions. Some questions are simply eliciting personal information about your skills and experiences. Some will clearly require research on your part to determine the best answers. Others will require you to give some careful thought and consideration to what you think the fire department is looking for. Don't assume you know what the "right" answer is. Research what others think about the policy, issue, or approach to a problem.

General

- What are your positive personality traits?
- What are your strengths?
- Why do you want to be a firefighter? (List from most important to least important.)
- Why do you want to be a firefighter for the city of _____?
- Why should we hire you?
- What computer skills do you have?
- Do you have a criminal record?
- What is your definition of stress?
- How do you deal with both everyday and unanticipated stress?
- What is one word your friends would use to describe you?
- What have you done to prepare for this interview?
- What have you done to prepare for a career in the fire service?
- Where do you see yourself in 5 years? 10 years?
- What is the most appealing aspect of being a firefighter?
- What is the least appealing aspect of being a firefighter?
- What do you consider to be your strongest asset? Your weakest?
- What activities have you been engaged in to improve / maintain your level of physical fitness?
- Why would you be a good firefighter?
- Why have you chosen to be a firefighter instead of a police officer or paramedic?
- What do you believe to be the most important attributes of a firefighter?

- What are your plans if you are not successful with your application?
- What physical demands do you expect to be placed on you during fire-fighting operations?
- Do you have any experience working in a team environment, and have you held any position of responsibility within the team / group?
- What do you believe are the characteristics of a well-balanced team?
- Define sexual harassment and give your opinions about the subject?
- What would you do if you were a witness to someone being sexually harassed?
- Honesty and integrity: Define them and why are they important in the fire service? (Use a dictionary.)
- The fire service has an equal opportunities policy. What do you understand by the term "equal opportunity"?
- The role of a firefighter involves working long hours at a time, night shifts, and public holidays. How will you cope with this disruption to your family life?
- How do you think you would cope when faced with casualties and / or fatalities at operational incidents?
- Pride and loyalty: Define them and why are they important in the fire service?
- You are __ years old. Why have you left it so long to apply for a career in the fire service?
- If you were asked to choose a hero, who would it be and why?
- What do you consider the definition of teamwork to be?
- What do you think you will dislike about this career?
- What does the term "diversity" mean to you?
- What three words describe your personality?
- How long have you wanted to be a firefighter?
- Do you have any problems with relocating?
- Do you have any family members who are firefighters?
- How many interviews with other departments have you had?
- Have you ever done a physical fitness test before? How did you do?

Work Habits

- What would your previous employer say about your work habits?
- What duties do you perform at your present job?
- Have you ever abused sick days at work?
- If your shift started at 8:30 a.m., what time do you think would be considered late?
- What type of person would you find most difficult to work with?
- Did you ever disobey an order? If so, what were the circumstances?
- How would you feel taking orders from somebody younger than yourself?—can you recount a time when you've been in this situation?
- How would you feel taking orders from a woman? Can you recount a time when this has been the case?
- If we were to contact your present or former employer, what would be one negative point they might raise about you?
- Attending operational incidents is an exciting and often rewarding experience; however, a significant part of a firefighter's day is spent on routine duties, such as cleaning and maintaining equipment. How do you think you would cope with the routine aspects of the job?
- Firefighters are required to use their hands in their work. What examples can you provide of any practical tasks you have done, either at home, at school, or at work?
- Our fire department cannot tolerate casual absences due to sickness or other reasons. Why do you believe the fire department is so strict with its absentee policy, and what are your thoughts on this matter?

Qualifications

- How can you relate your education and work experience to firefighting?
- Do you think you are qualified enough to be a firefighter?
- What skills do you have that would be a benefit to the fire department?
- What fire-related training do you have?
- What responsibilities do you have in either your job or domestic activities?
- Have you any experience in communicating with the general public?

Give some examples. What do you consider the most important factors in being able to communicate clearly with the public?

- One aspect of a firefighter's role is to produce written reports. Would this present any problems or concerns?
- How do you arrive at making difficult decisions?
- What practical medical experience do you have?
- What was your reasoning for attending _____ College?
- Have you ever handled an emergency situation?
- Have you ever experienced a dangerous situation. If so, how did you cope? What were your thoughts and what were your priorities?
- What did you like about being a volunteer firefighter?

Fire Department Knowledge

- What do you like about our city?
- What do you like about our fire department?
- How many calls do we run annually?
- What medical training is required to work here?
- What roles do our firefighters play in the community?
- What is our chief's name?
- Have you ever visited any of our fire stations?
- Do you know how many fire stations we have?
- Under what circumstances would you disobey an order?
- What special teams do we have?
- If hired, what contributions could you make to our department?
- What is the typical daily routine in a firehouse?
- What will you do with your spare time while on duty?
- What duties does a firefighter perform?
- What is the most essential duty a firefighter performs?
- What other special services does the fire department offer to the public?
- How could you help maintain good relations around the firehouse?
- What single trait must every firefighter possess?
- What do you think the mission statement should be for our department?
- In what direction do you see the fire service going?

Behavioral

These questions are designed to provide insights into an applicant's abilities and skills in such matters as negotiation, persuasiveness, teamwork, communication, decision making, problem solving, planning, organizing, and coping with pressure.

- Tell us about a time when you had to deal with an angry or hostile person (e.g., a customer, employer, co-worker, friend).
- Tell us about a high-stress situation you have experienced.
- Tell us about a situation where you helped someone you didn't know.
- Tell us about a situation where you were responsible or accountable for someone else.
- How would you react to unwanted attentions from a co-worker (e.g., bullying, name calling, initiation ceremonies)?
- Can you tell us about a time when you told a lie and had to, or chose to, confess?
- Tell us about a time when you had to exercise self-discipline.
- Can you explain the difference between discipline and self-discipline?
- Discipline is considered to be very important in the fire service. Why do you think fire departments promote discipline?
- Can you tell us about a time when you've been caught in an embarrassing situation? How did you handle it?
- Tell us about a time when your positive attitude influenced others to be positive or motivated.
- Confidentiality is critical in the fire service. Why do you think it is so critical?
- How would you handle a conflict with a co-worker?
- Tell us what you would do if you caught a co-worker stealing?
- Tell us what you would do if you witnessed a female co-worker being sexually harassed by a male co-worker.
- Tell us about a time when you influenced someone in a positive way.
- Tell us about a time when you had to stand up for someone.
- How would you handle working with a co-worker who has a negative attitude?

- Give us an example of when you took the initiative at work.
- Tell us in your own words the meaning and significance of rank and role.
- Tell us about a time you disagreed with your supervisor and how you handled it.
- Tell us about a situation in which you worked with others in order to accomplish a common goal.
- Tell us about a time you experienced taking responsibility for your actions.
- Tell us about a time when you were involved in an emergency situation. Describe the situation and how you handled it.

Hypothetical Situations

These questions are used to assess your problem-solving skills, sometimes in situations that require a very quick reaction.

- We hired you, and your home town soon announces they are hiring. What would you do?
- You are at a fire call and your captain asks you to climb a ladder you know is unsafe. What would you do?
- A co-worker is making racist comments and it's bothering you or another co-worker. What would you do or say?
- Your captain takes you aside and you notice he has been drinking. What would you do?
- You are walking past a house, smell gas, and notice what looks like someone unconscious in the house. What would you do?
- You are evacuating an entire floor of an apartment building because of a small fire on the floor below. You encounter a hostile occupant who refuses to leave his apartment. How would you handle this situation?
- You have just found out that a co-worker has been accused of committing a criminal offense. What would you do?
- While you are exiting a building, you encounter an individual who appears to be unconscious on the sidewalk. What is the most important thing you would do?

- You are at an off-duty party and notice your captain smoking drugs. What would you say or do?
- You have just returned to the fire hall from a life-threatening incident when you notice a relative of the victim waiting there. How would you approach this situation.
- You arrive at a structure fire, which is puffing smoke from its windows and doors and has been burning for a while. Your captain instructs you to ventilate the roof. What would you do?
- You arrive at a structure fire and notice that the building looks abandoned and that the roof is showing signs that it is about to collapse. Your captain orders you to advance a hose line to the front door and prepare to enter the structure. What would you do?

Practical—Fire

- What is a flashover?
- What is a backdraft?
- What are the signs of a building about to collapse?
- What are the types of building collapses?
- What are the classifications of fire and the associated letters. Explain how each extinguishes fire?
- Describe the different methods of heat transfer.
- Explain the fire tetrahedron.
- Name the classifications of building constructions.
- What are the steps you would take at the door before entering a structure on fire?
- Describe the difference between a wye and a siamese.
- What are the different categories of skin burns? Explain.
- Explain the chain of command.
- What are the different stages of fire?

Practical—Medical

- What does ABC stand for in CPR?
- What is the rate of respiration for an adult, a child, and an infant?
- What is the average resting heart rate for an adult, a child, and an infant?
- Name three different types of fractures.
- What are the indications of a head injury?
- What does OPA stand for when it comes to human airways?
- What is the number of chest compressions for an adult during single rescuer CPR?
- How many breaths per second should there be during CPR for a child?
- How do you treat a person suffering from heat stroke?
- What does BVM stand for?

FIREFIGHTER GOES THE DISTANCE DRIVING RIG ACROSS CANADA FOR DONATED GEAR.

A Toronto firefighter is driving a rig across Canada today in hopes of filling it with firefighting equipment for poor countries. So far Ron Kyle, of Barrie, has collected about $200,000 worth of firefighting equipment from Toronto Fire Services, Central York Services, and now he's on his way to the Calgary Fire Department to pick up another load.

— Tracy McLaughlin, *Toronto Sun*, September 3, 2004

PRACTICING THE INTERVIEW

The best preparation for the interview is to practice, using mock interviews with someone who can be helpful. Have someone you know ask you questions that you can practice answering in an interview setting. (You may even be able to convince someone from your city's human resources department to help you practice.) Use the interview questions provided in this chapter as a guide. By practicing, your confidence will increase because you will be better prepared (there'll be fewer pauses with "Uh") and you will then feel more at ease.

Candidates spend many hours increasing their fitness level, volunteering, and taking courses, but may only spend a short time preparing themselves for the interview. This is a mistake. If you go into an interview just "winging it"—you will not do as well as you could have, had you been better prepared.

No doubt, you will be nervous and we all know what nerves can do to us. I'm not suggesting that you memorize your answers word for word. However, be sure to review the example interview questions and prepare answers. Review them again the night before the interview. For example, you may know "why you want to become a firefighter," but do you know "why you would make a good firefighter"? If you stumble in your response to the interview committee, then they won't get to learn what you have to offer, and if you hesitate, you may not seem sincere. By being prepared, you will impress the committee with your confidence. Don't be in a hurry; they want you to succeed.

WHAT TO WEAR

Your interview is a very important part of the entire recruitment process—perhaps the most important. Clothing for men should be conservative business attire, such as a suit, dress shirt, and tie; don't wear jeans and sneakers. And make sure your clothes are clean and freshly pressed. Women should wear conservative business clothing as well. Hair should be neatly groomed, and applicants with long hair should tie their hair back or up for a more professional look. This is a crucial moment in your life. You have already spent perhaps thousands of dollars on courses, so spend a little more on nice clothing. Proper business attire and a confident professional attitude will go a long way to convincing the committee that you really want the job and are taking the interview very seriously.

HOW TO APPROACH THE INTERVIEW

Try to imagine what it is like to be one of the interviewers on the committee. It has been a long week, they have interviewed hundreds of candidates, and much to their relief, in walks the final applicant—and it's you. You will make a good first impression if you are well-dressed, but they will be inclined to think that you will be like everyone else interviewed to this point. It is now your job to make yourself stand out. Studies have shown that you have thirty seconds to catch the interviewer's attention, and if you don't within that period of time, that person will put his or her mind on autopilot until you are done your interview. A way to combat this is by not overdoing your opening statement. Fire departments like to use the question "Tell us about yourself." This is supposed to be an easy icebreaker to help you relax and adjust to the environment. Your answer should be no more than one minute in length. Talk about your family or hobbies, and keep it simple. You don't want to overdo it. If you spend 14 minutes of a 20-minute interview talking about your hobbies, then how is it possible to get through the 20 remaining questions in 6 minutes?

Look at the interview from the fire department's perspective. A major decision is being made here. The department is hiring for the long term. Unlike other major decisions that people make, like buying a house, or taking a job, this decision cannot be reversed. You can sell your house or quit your job, but when a fire department hires you, they are making a 25- to 30-year commitment. The interviewers want to assure themselves that you are who you say you are, and that you believe in what you say you believe in. The interview provides a very short period of time to determine who is best suited for the job—and I'm not talking about qualifications because if you weren't qualified, you wouldn't have got the interview in the first place. I'm talking about character. Do you possess the traits or qualities that the fire service is built upon—honor, integrity, adaptability, discipline, excellent people skills, determination? So keep these traits or qualities in mind when you are preparing answers to questions you could be asked.

LAST-MINUTE ADVICE

- Get a good sleep the night before.

- Stay positive and keep reassuring yourself that you're doing fine.

- Keep doubts to yourself.

- Try to think of a prior rejection as a positive thing—you have learned from your mistakes.

- Remain confident—you are well-prepared.

- It's expected that you'll be nervous, so don't sweat it. Concentrate on your answers, and not on how you feel.

- Combat nervousness by deep breathing before the interview (sparingly). This will help relax your body, which will then lead to a calmer mind.

- Do only a light review when preparing the night before. It's all you need if you have spent time preparing properly in the preceding weeks.

CHAPTER TWENTY-TWO

HOW TO SURVIVE THE INTERVIEW

PRIOR TO THE INTERVIEW

Do

- Prepare a list of your qualifications, experiences, and personality traits. Refer to this list when selling yourself in the interview.
- In case you have to clarify a section of your résumé, make sure you know it inside out so you can answer a question without hesitating.
- Do thorough research. Make sure you have important information about the fire department, including its fundraising efforts and community involvement. Write everything down in an organized manner so it is readily available.
- Know your strengths. Write down experiences you have that match each qualification listed in the posting.
- Know your weaknesses, and explain your efforts to improve them. Fire departments like candidates who can acknowledge their weaknesses. But try to describe weaknesses that are more positive than negative. For example, "My friends say that I am too serious and spend too many weekends taking courses, and not enough time having fun." This is much better than saying

"I have a difficult time following instructions when I am excited." The first example is relatively harmless; the second example would be a major concern to the fire department.

Don't

- Don't bring copies of your résumé to the interview. If you want everyone on the panel to have a copy, arrive well in advance and give them to an administrative assistant to distribute to the panel. Become a member of www.becomingafirefighter.com and have your résumé and cover letter reviewed.

DURING THE INTERVIEW

Do

- Concentrate on general questions the interviewers might ask, such as "Why do you want to be a firefighter?" "What can you offer?" and "What are your strengths and weaknesses?"
- Give a firm handshake and smile when introductions are made. Look each person in the eye and don't look away. You should show that you are genuinely interested in meeting them.
- Fold your hands in your lap. This will help you maintain proper posture during the interview. Sit up straight and look attentive so that you appear keenly engaged.
- If you have problems with sweaty palms, expose them to the air, or place them on your knees.
- Be polite and courteous.
- Change the pitch of your voice rather speaking in a monotone, which can sound boring.
- Listen to the whole question and let it sink in before you attempt to answer.
- If you didn't fully understand the question the first time, ask the interviewer to please repeat the question. Not only will this give you more time to answer the question, but now you have heard the question in two dif-

ferent ways. But be careful not to take too much time—you have only twenty minutes or so.

- Ask a question if you need clarification. The interviewers' question may be, "You see your partner putting a tool in his pocket. What do you do?" Ask yourself, "What are they getting at?" Are they telling you that your partner is stealing? Don't assume anything—ask a question, such as "Is the tool his?" If they reply, "No" then you've already scored points. I would then say, "Stealing is wrong, and if he doesn't put it back, I would say, 'Let's go together and talk to our captain, because removing tools or equipment without approval is stealing, and stealing is against the law and the rules of the department.'"
- Watch your facial expressions and body language. Don't chew gum, crack your knuckles, tap your foot, jiggle your foot, or show impatience in any way.
- Watch your language. Don't use profanity—you're not with your friends.
- Be friendly to office staff, such as a receptionist, or anyone else you meet before or after the interview.
- Maintain eye contact with everyone on the committee. However, looking away or at your notes to ponder a question for a moment is acceptable.

Don't

- Don't use first names (unless requested). Using "Sir" or "Ma'am" demonstrates that you are polite and respectful.
- Don't smoke (even if invited).
- Don't sit down until invited.
- Don't show anxiety or boredom, and don't look at your watch.
- Don't always refer to your notes.
- Don't use a flashy pen or writing implement that could distract you or the interviewers.
- Don't provide unnecessary (too much) information.
- Don't ask about salary, vacation time, or benefits—bad timing! You should already know this information by researching properly.
- Do not mumble or ramble. Speak loudly and clearly. Answer intelligently, but be brief.

TIP!

During the interview, it is best not to ask too many questions and to simply respond to the questions posed, but it is a good idea to make a closing statement to underline your qualifications and to reinforce your enthusiasm for the job.

END OF THE INTERVIEW

Do

- When the interview is coming to an end, thank them for their time, regardless of how you feel the interview went. Always appear confident. Don't show disappointment.
- Send a "thank you" letter addressed to each individual of the interview committee within a day of the interview. Showing consideration and keeping a connection can affect your chances of being recruited. Ensure that you have the correct spelling of the names of all committee members.
- Always leave in the same polite and confident manner as you entered.

Don't

- Don't indicate that you feel discouraged because you believe the interview did not go well; keep it to yourself, even after you've exited the building. You never know who is watching—never express anger publicly.
- Don't appear to be overly confident if you felt the interview went well; appear humble and grateful. Keep it to yourself—it shows maturity.

If by chance you have a juvenile offence that you can't have removed from your record, make sure you are upfront about it in an interview or if ever asked about it. The bottom line is, if a fire department is interested in you, they will find out about the offence, so you might as well be frank. This will also demonstrate that you can admit your mistakes and that you are a trustworthy person.

CHAPTER TWENTY-THREE

RÉSUMÉ AND ACCOMPANYING LETTERS

COVER LETTER

A cover letter should always accompany your résumé and is used whenever mailing, e-mailing, faxing, or personally delivering your résumé. You can submit your résumé and cover letter to a fire department whether or not they are actively recruiting.

A cover letter is designed to give a brief indication of the most relevant skills you have, and if you are responding to a job ad, it should address the skills mentioned in the ad, without copying it too closely. Cover letters also give you an opportunity to demonstrate your writing skills. Studies have shown that well-written cover letters enhance the chance of a résumé being read by up to 65 percent. The cover letter must be able to stand on its own, even though it always accompanies a résumé. Do not send it hand written. Always keep a copy.

- Format your cover letter to resemble a typical letter you would send any employer. All information such as your name and address and the recipient's name and address should be aligned on the top left margin of the page. For example,

John Doe
224 Ernest Ave.
London, ON
W3E 4W3
(555) 000-0000

_____ (Recipient's Name)
Human Resources Department
City of Toronto
City Hall, 1st Floor
500 Front Street
Toronto, ON
N6V 3T3

- The actual letter should consist of short paragraphs and should never be more than one page in length, including your signature.
- Always find a specific individual you can address your cover letter to, rather than a general recipient such as the "Ottawa Fire Department" or "Human Resources Department." Call and find out who is chairing the hiring committee.
- The cover letter should always be dated.
- Use upper and lower case letters. NEVER USE ALL CAPITALS.

- Don't state the obvious —that you are interested in becoming a firefighter.
- It is not necessary to list every qualification you have. Remember, it is only an overview of your qualifications. Always list the most important qualifications you have first.
- Each fire department you apply to should receive a personalized cover letter. If a fire department lists in its recruitment announcement that they are looking for recruits that possess a certain quality or qualification—and you have it—say so in your covering letter first.
- Don't send a cover letter that contains any typos, misspellings, incorrect grammar or punctuation, smudges, or whiteout. It must be perfect. Have someone else proofread it before sending.
- Use simple language, uncomplicated sentence structure and short sentences.
- Use short paragraphs (four to five lines).
- Always sign the letter with ink. Your name should be typed beneath your signature.
- Use good quality paper.
- Only use black ink, no color.
- Don't include photographs, pictures, or diagrams.
- Don't overuse the word "I."
- Avoid any negativity.
- Don't use slang or profanity, and avoid clichés.

FIGURE 23.1 • SAMPLE COVER LETTER

John Doe
224 Ernest Ave.
London, ON
W3E 4W3
(555) 000-0000

September 04, 2006

_____ (Recipient's Name)
Human Resources Department
City of Toronto
City Hall, 1st Floor
500 Front Street
Toronto, ON
N6V 3T3

Dear_____,

Thank you for the opportunity of applying for the Probationary Firefighter position. I am confident in my ability to achieve the standards required by the Toronto Fire Department.

I have recently completed NFPA 1001 (standard for professional qualifications) and NFPA 472 (hazardous materials operations) at _____ Fire School in _____ (province/state). In addition to graduating from _____ Fire School, I have successfully completed_____ College's paramedic program. Since graduating, I have been a certified paramedic for ____ years.

I feel that my education and experience coupled with my strong desire to work in the fire services make me an ideal candidate. I am an outgoing, optimistic individual with a positive attitude who would be a great addition to any of your teams.

I would greatly appreciate an opportunity to discuss your requirements and my qualifications. I am available for an interview at your convenience.

Sincerely,

John Doe

John Doe

RÉSUMÉS

There is only one purpose of your résumé: to convince the fire department to give you an interview. A résumé is an advertisement; nothing more, nothing less. A great résumé doesn't just tell the fire department what you've accomplished, but makes the same assertion that all good advertisements do: if they buy this product (you), they will get these specific benefits. It must present you in the best light and convey that you have what it takes to be a successful probationary firefighter.

It is vital that your résumé is pleasing to the eye and easy to follow so that the recruiter is enticed into picking it up and reading it thoroughly. The recruiter should want to meet and learn more about you. Take the time to ensure your résumé is the best it can be.

Résumé Writing Tips

- Be sure to limit the amount of pages used for your résumé. Two pages is ideal, but three or four pages is fine if absolutely necessary to convey your skills and qualifications.
- Always put your name, address, postal code, phone number, e-mail address, and page number at the top of each page in case the pages get separated.
- Always list education, volunteer work, and experience in reverse chronological order (most current to least current).
- Be sure to use boldface for your headings, so they will stand out.
- Do not use italics in your résumé.
- Use upper and lower case letters—NOT ALL CAPITALS—except for headings.
- Do not list months or days when listing courses or jobs—only the year(s).
- Refer to your current position in the present tense. All other entries should be written in the past tense.
- Use an easy-to-read font style and nothing smaller than a 12-point font.
- Use at least 1.5-inch (4-centimeter) margins on each side.
- Use point form when describing job duties, education, background, and so on. Include very brief descriptions.

- Use the following order for your résumé headings:

  PROFILE or SUMMARY OF QUALIFICATIONS
  FIRE TRAINING or FIRE EXPERIENCE
  EDUCATION
  EMPLOYMENT
  SKILLS (licenses and certifications listed separately)
  VOLUNTEER EXPERIENCE
  REFERENCES

  Note: Every piece of information on your résumé should fall under one of the above categories.

- Don't put more information on your résumé than is necessary. If it is crowded with information, it is difficult to draw attention to the points you wish the reader to see.
- Be organized, relevant, accurate, and concise but with sufficient detail.
- Include the year for each entry (use "present" if currently in the position or course rather than the current year, i.e., 2007–present). This will eliminate confusion as to whether you are still employed in that position.
- Be sure to have consistency between the cover letter and the résumé. If you mention a skill in your cover letter, make sure you include it in your résumé.
- Use only black ink—no color.
- For employment, identify the job title in boldface and list it first. Keep in mind the job title is more relevant than the company name, although the company name must be there.
- Be sure to review the minimum qualifications list to determine exactly which skills are important to the fire department, and clearly identify those skills in your résumé.
- Don't forget to include any co-op placements under work experience.
- Before you put a pen to paper, review other firefighter résumés and compare them with each other. Pick out things you like about them—it could be the format, the layout, or the language in the text.
- For each job, try to convey how your skills, responsibilities, and accomplishments demonstrate how or why they are relevant and transferable.

- Do not make assumptions about the recruiters' knowledge. Nothing is obvious. They should not be trying to decipher which skills you may or may not possess. Be very clear.
- Never include your picture on your résumé or separately submit a photograph. Age, ethnicity, and appearance are not things you want to be judged on the basis of.
- Never include your hobbies, interests, or family information. These do not belong on a résumé! However, if you have an experience that you think will be relevant, for example, that you once played professional hockey, list this information under the "Profile" category.
- You should also color photocopy all of your course certificates to submit with your résumé. It is acceptable to use a scanner to do this. It is a minor thing, but you would be surprised how many applicants photocopy their certificates in black and white. Color copies look that much better!
- If you attended a fire college, photocopy and submit all your transcripts with your résumé package, along with any awards. If your grades were good, this is helpful. The transcripts should be located in your résumé package between your firefighting diploma and other certificates.

The Seven Deadly Résumé Sins

1. **Never lie.**
   Aside from the moral implications, if you are considered for employment and your fabrications are found out, you will obviously not get the job, you will have damaged your chances at getting hired in the future, and, more importantly, you will have damaged your reputation generally.

2. **Don't use the word "résumé" on your résumé.**
   The recruiters can clearly see that they are reading a résumé.

3. **Never include salary information or expectations.**
   You can determine your potential pay scale by contacting the city's human resources department. Firefighters are city employees, and taxpayers are entitled to know the salary scale of city employees.

4. **Don't attach job references or testimonials.**
   Usually at the end of the resume it will say, "References available upon request." This is sufficient. You may include your references and any testimonials you have in your application package.

5. **Never include personal data and photographs.**
   Details about your marital status, age, height, weight, and so on are not important and are only invitations for discrimination.

6. **Don't describe your hobbies and personality traits.**
   I like hockey, hiking, swimming etc ... Instead, the recruiters will gain an understanding of your personality from the interview.

7. **Don't copy someone else's résumé. Be original and creative.**
   Start your résumé from scratch. You can use the sample résumés in this book or others that you find as a template, and for ideas for ways to express your skills and strengths, but be creative and original. Your résumé is an expression of who you truly are.

See the sample résumés in Figures 23.2 and 23.3.

FIGURE 23.2 • SAMPLE RÉSUMÉ

JOHN DOE

224 Ernest Ave. / London, Ontario / W3E 4W3 / (555) 000-0000 / jdoe@email.com

PROFILE

Summary of Qualifications
- Played professional hockey—Calgary Flames
- Excellent interpersonal skills with team-building qualities
- Able to resolve conflicts while remaining sensitive to others' feelings
- Self-motivated learner, constantly striving for improvement, constant reader
- Responds positively to problems while working on creative solutions
- Remains calm under pressure to overcome conflicts and problems
- Mature lifestyle with focus on health and safe work practices
- Comfortable and experienced with work in adverse environments

FIRE TRAINING

NFPA 1001—Fire Fighter Level I and II	2009
Fire Etc	Vermilion, AB
NFPA 472—Hazardous Materials Operations Level	2009
Fire Etc	Vermilion, AB
NFPA 472—Hazardous Materials Awareness Level	2009
Fire Etc	Vermilion, AB

EDUCATION

Alarm Technician	2008–2009
Alarm College	London, ON
Sprinkler Technician	2007–2008
Sprinkler College	London, ON
Secondary School Diploma	1999–2003
London District Collegiate Institute	London, ON

FIGURE 23.2 • SAMPLE RÉSUMÉ

224 Ernest Ave. / London, Ontario / W3E 4W3 / (555) 000-0000 / jdoe@email.com

EMPLOYMENT

Patient Transfer Technician	2008–present
Patient Transfer	London, ON
Electrician Apprentice	2007–2008
Wayne's Electrical Services	London, ON
Quality Assurance Inspector	2003–2007
Boltons Manufacturing	St. Thomas, ON
Laborer	2003
City of London	London, ON

RELEVANT SKILLS

LICENSES:
- B-Z License—Fanshawe College   2010

CERTIFICATES:
- WHMIS—Patient Transfer   2010
- Basic Trauma Life Support—Pediatric—Toronto EMS   2010
- Basic Trauma Life Support—Toronto EMS   2009
- CPR Level C—Life Support Services   2009
- Semi-Automated External Defibrillation—Life Support Services   2008
- Medical First Responders—Canadian Red Cross   2008
- Standard First Aid—Canadian Red Cross   2008

VOLUNTEER EXPERIENCE

- Emergency Responder, Red Cross, Hamilton Branch   Present
- Firefighter Combat Challenge, Hamilton, ON   2010
- St. John Ambulance, London, ON   2009

References available upon request

FIGURE 23.3 • SAMPLE RÉSUMÉ

MARTIN DOE   Pg. 1 of 2   (Sample # 2)

142 Lundy Lane • London, Ontario • N5R 8L7 • (555) 123-4567 • mdoe@email.com

SUMMARY OF QUALIFICATIONS

- Dependable team player
- Willing to assume all levels of responsibility
- Excellent communication and interpersonal skills
- Thorough understanding of electrical and mechanical components
- Ability to organize and prioritize projects in a fast-paced environment
- Excellent manual skills and technical aptitude
- Conditioned to working in extreme cold and hot ambient environments

FIRE EXPERIENCE

Volunteer Firefighter	2008–present
Penticton Fire Department	Penticton, ON
Pre-Service Firefighter Certificate	2009–2010
Lambton College	Sarnia, ON
•Co-op, St. Thomas Fire Department (360 Hours)	2010
Fire Science and Technology	2008
Lambton College	Sarnia, ON
• Co-op, Toronto Fire Department (420 hours)	2009
• Co-op, Sarnia Fire Department (460 hours)	2009

EDUCATION

Refrigeration and Air Conditioning Program	2001–2003
Mohawk College	Hamilton, ON
Ontario Secondary School Diploma	1997–2001
Sir Duke Secondary School	London, ON

FIGURE 23.3 • SAMPLE RÉSUMÉ

MARTIN DOE   Pg. 2 of 2   (Sample # 2)

142 Lundy Lane • London, Ontario • N5R 8L7 • (555) 123-4567 • mdoe@email.com

EMPLOYMENT

Heating, Ventilation, Air Conditioning,   2003–present
   and Refrigeration Mechanic
   Petrillo's Air Conditioning and Heating Company   London, Ontario
- Completed five-year apprenticeship for air conditioning and refrigeration
- Health and safety representative
- Respond to natural gas and propane leaks as well as carbon monoxide alarms
- Installation, repair, and maintenance of positive displacement pumps, centrifugal pumps, motors, air conditioning, heating and refrigeration equipment
- Welding and soldering (arc, brazing, oxy-acetylene, acetylene)

RELEVANT SKILLS

Licenses:
- Air Conditioning and Refrigeration Mechanic (Inter-provincial)   2008
- Gas Technician II   2006–2009
- Class B Driver's License (includes C, D, E, F, and G)   2009

Certifications:
- Automated External Defibrillation   2010
- BTLS Level 1   2010
- CPR and First Aid Instructor   2006
- Air Brake Endorsement   2010

VOLUNTEER EXPERIENCE

- St. John Ambulance Brigade (three to four hours/week)   2009–Present
- Big Brothers of London   2009
- Assistant Coach for Tykes Baseball Team   2000

References available upon request

Helpful Words and Phrases

Add descriptive words and phrases like the ones listed below when explaining the items on your résumé.

- Highly organized and dedicated, with a positive attitude
- Able to handle multiple assignments under pressure
- Excellent written, oral, and interpersonal skills
- Thrives on working in a challenging environment
- Widely recognized as an excellent care provider and patient advocate
- Superior accuracy in patient history, charting, and other documentation
- Dependable team player
- Willing to assume all levels of responsibility
- Thorough understanding of electrical and mechanical systems and components
- Able to organize and prioritize projects in a fast-paced environment
- Excellent manual skills and technical aptitude
- Conditioned to working in extreme cold and hot ambient environments
- Excellent computer skills
- Works well with others in order to achieve a common objective
- Strongly able to motivate and encourage others
- Able to see opportunities and to set and achieve goals
- Thinks things through logically to determine key issues
- Able to handle change and adapt to new situations
- Relates well to others and establishes good working relationships
- Competent understanding of numerical data, statistics, and graphs
- Excellent interpersonal skills with team-building qualities
- Able to resolve conflicts while remaining sensitive to others' feelings
- Self-motivated learner, constantly striving for self improvement
- Adaptable to change and diversity
- Responds positively to problems while working on creative solutions
- Remains calm under pressure to overcome conflicts and problems
- Mature lifestyle with focus on health and safe work practices
- Able to work in adverse environments
- Search blog.emurse.com for great "power words" to incorporate into your résumé

Organize Your Data

Prior to starting your résumé, use the outline below in Figure 23.4 to help you collect and organize the information needed. Be careful to confirm the proper names and the correct locations and dates for all entries.

FIGURE 23.4 • ORGANIZING YOUR DATA

FIRE TRAINING: List of Accomplishments

Course/Experience: _____

Location: _____ Date: _____

Course/Experience: _____

Location: _____ Date: _____

Course/Experience: _____ _____

Location: _____ Date: _____

EDUCATION: List of Accomplishments

Course: _____

Location: _____ Date: _____

Course: _____

Location: _____ Date: _____

Course: _____

Location: _____ Date: _____

Course: _____

Location: _____ Date: _____

EMPLOYMENT: List of Accomplishments

Position: _____

Company: _____

Location: _____ Date: _____

Position: _____

Company: _____

Location: _____ Date: _____

Position: _____

Company: _____

Location: _____ Date: _____

Position: _____

Company: _____

Location: _____ Date: _____

RELEVANT SKILLS: List of Accomplishments

Course/Experience: _____

_____

Location: _____ Date: _____

Course/Experience: _____

_____

Location: _____ Date: _____

Course/Experience: _____

_____

Location: _____ Date: _____

VOLUNTEER EXPERIENCE: List of Accomplishments

Organization/Involvement: _____

_____

Location: _____ Date: _____

Organization/Involvement: _____

_____

Location: _____ Date: _____

Organization/Involvement: _____

_____

Location: _____ Date: _____

Organization/Involvement: _____

_____

Location: _____ Date: _____

REFERENCES

Give a lot of thought to whom you select as your references. It would be a terrible shame if you had impressed the selection committee up to this point, only to have them underwhelmed by your references. This is one area where you have control, so ensure you choose wisely to reinforce the positive impression that you have given so far. The best advice I have is, if possible, choose people who have *offered* to be a character reference for you. They will certainly say positive things about you and will convey enthusiasm for you and your qualities.

If you are concerned that you have weak references, take the time to create good ones. Get out and meet new people, visit fire halls, and talk with all the staff. It's not hard to meet and get to know new people that have the same interests as you do. Many people in the fire service are happy to take someone under their wing and help out. Most firefighters remember how hard it was to get a job, and they'll enjoy being your mentor or friend. If someone does take the time and energy to help you out, make an extra effort to thank the person promptly and keep in touch. In fact, you should thank all references, whether you get the job or not.

If you are asked to provide references, prepare an information sheet similar to the one below in Figure 23.5.

FIGURE 23.5 • SAMPLE REFERENCE LETTER

JOHN DOE   Pg. 1 of 1   (Sample # 1)

224 Ernest Ave. / London, Ontario / W3E 4W3 / (555) 000-0000 / jdoe@email.com

---

REFERENCES

Name of reference
Phone #
Title (Employer)
Address
Postal Code

Name of reference
Phone #
Title (Firefighter)
Address
Postal Code

Name of reference
Phone #
Title (Co-worker)
Address
Postal Code

FOLLOW-UP LETTERS

You may find this surprising, but the follow-up letter is extremely important. Consider this scenario: The fire chief and the other committee members are preparing to make their decision. You're one of the applicants in the running for a position and they are having a tough time making a decision. That morning, your follow-up letter arrives. It is professionally prepared, the tone is pleasant and enthusiastic, and you reiterate your interest in the position. Guess who most likely gets to the top of the job offer list?

FIREFIGHTERS HONORED FOR BRAVERY

The Ontario government recognized two firefighters and three police officers for bravery during a ceremony held Feb. 2 at Queen's Park.

James K. Bartleman, Lieutenant Governor of Ontario, was joined by Community Safety and Correctional Services Minister Monte Kwinter to present the medals. These medals are the province's highest honor in recognition of firefighters and police officers whose actions demonstrate outstanding courage and bravery in the line of duty.

"The heroic actions of firefighters and police officers remind us of the risks they face every day," said Bartleman. "We owe them our gratitude and our thanks for all they do to keep us safe."

The recipients of the 2005 Ontario Medal for firefighter bravery are Capt. Ralph Novel of the Mississauga Fire and Emergency Services, and Lieut. Mark Smith of the Sarnia Fire Rescue Services.

On December 21, 2004, Capt. Ralph Noble was walking his dog in a park area near Lake Ontario when he heard a faint cry for help. He saw a woman in the water about 35 meters from the shore. Noble caught the attention of a bystander and asked her to call 9-1-1. He then took off his jacket, unleashed his dog, and entered the icy waters to rescue the woman by throwing the leash towards her. Despite the victim's advanced stage of hypothermia, she held on to the leash and Noble pulled her in, saving her life.

An independent body of citizens representing all regions of Ontario determine medal recipients. A total of 163 Ontario Medals for Firefighter Bravery have been awarded since 1776.

Canadian Firefighter and *EMS Quarterly*, April 2006, p. 8.

CHAPTER TWENTY-FOUR

MAKING THE SHORT LIST

A hiring pool is a short list that consists of qualified applicants who have been accepted during a recruitment process. These applicants are usually rated or scored to determine the order in which they will be hired. Fire departments rarely disclose your current position on the short list. Applicants on this list are expected to continue to take courses and improve themselves in order to maintain their position on the list. Failure to improve your skills and experiences during this stage will only serve to be an advantage to other applicants. Once you have made the short list, my advice is to take a week or two, maybe even a month, to get your thoughts together and plan your next step. Then, enroll in new courses and engage in new volunteer or other experiences so you can remind the fire department of why you made the short list. Continue to improve your skills until the day you are recruited. This goes a long way to showing commitment and determination.

Keep in mind that regardless of your ranking on the short list, you may be hired sooner than you think. You may be ranked fourth, let's say, but other applicants on the list who are ranked higher than you may well have already been hired by other fire departments, or may refuse the job offer due to the fact that they're going through the recruitment process for the fire department in their home town. The short list can stay active for one to two years, depending on how many applicants remain on the list. Short lists will collapse (those on them get hired elsewhere), and then fire departments will host another recruitment process and create a new hiring pool to fill vacant positions.

It is vital that you continue to update your personal file at the fire department. This gives you a legitimate reason to visit the fire hall again and you can then visit with the chief, deputy chief, and others firefighters. When updating your personal file at the fire department, make sure you take the same professional approach as you did with your résumé. It is recommended that you follow the same organizational format as you did for your résumé. It's important that your name and contact information is included as well as the updated material, including course taken or experience gained, location, and date. If you intend to update your personnel file on the day of your interview, be sure to give your information to the administrative assistant prior to the interview and ask that they deliver the material to the interview committee. This is a more professional approach than handing it out at the interview, and gives the committee time to review the information before the interview starts.

Courage is what it takes to stand up and speak;
Courage is also what it takes to sit down and listen.
— Carl Hermann Voss

FIGURE 24.1 • SAMPLE RÉSUMÉ UPDATE LETTER

JOHN DOE

224 Ernest Ave., / London, Ontario / W3E 4W3 / (555) 123-4567 / jdoe@email.com

FIRE TRAINING

Ice Rescue   **2011**
   Fire College   Toronto, ON

Confined Space   **2011**
   Fire College   Sarnia, ON

VOLUNTEER EXPERIENCE

Food Bank   **2011**
                                                          London, ON

CHAPTER TWENTY-FIVE

VISITING THE FIRE HALL

Visiting a fire hall can have a great impact on where and when you get hired. You should try to visit every four to six months, if possible. Your objective is simply to have the fire department put a face to your name and to make them aware that you're interested and live in the general area.

So, what should you talk about once you've introduced (or re-introduced) yourself? Ask simple questions, such as "Has it been busy today?" "How long have you worked here?" and "Tell me about the last fire you attended?" These types of questions are easy to answer and help to create a relaxed atmosphere. Don't bombard the firefighters with technical questions. And don't talk about yourself too much, as you don't want to appear to be bragging. Remember, the word "I" is not as important to the listener as the word "you." This shows you are a team player. Limit the length of the visit to ensure the quality and impact of it.

During your visits, make sure you exhibit a positive attitude. Leave any negative energy at home. The last thing you want to do is vent your frustrations and tell them about how hard it is to get hired or how you should have received an interview in the last recruitment round. If the subject of your job efforts does arise, say something positive, such as, "It must have not been my turn to get hired. I know I'll get my chance; I just have to be patient. Do you have any ideas how I can improve my chances?"

Don't visit too often or you may be considered a nuisance. However, you do want them to continue to know that you're eager and part of the community. It would be nice to show your appreciation for the time they take with you by bringing in a tasty treat for the on-duty staff. Everyone will be wondering where the treats came from, and your name is going to be mentioned. This is exactly what you want. As far as time of day to visit, you should refrain from visiting during lunch or shift changes. Most fire halls will be extremely receptive to you regardless of when you happen to visit. If you are uncertain, you could always call ahead and set-up an appointment for the visit.

In order to leave a good impression on your visit to the fire hall, the following are some key points to consider.

Appearance—Casual clothes are appropriate, such as jeans, shorts, and T-shirts. However, avoid wearing shirts that have suggestive or controversial statements on them. You should be clean shaven, well groomed, and wearing little or no makeup. Also avoid strong scents. If you have piercings, you may want to remove them; if you have tattoos, you may want to conceal them. Although this is certainly not necessary, some people are off-put by these body "embellishments."

Confidence—Be sure to have a firm handshake and maintain eye contact. Ideally, you should shake hands greeting and leaving the hall.

Manners—Make sure you demonstrate proper manners throughout your entire visit. Never chew gum or wear sunglasses when greeting someone or engaging in conversation. It's important they see your eyes to be convinced of your sincerity. A simple "thank you for your time" will suffice when you leave, and be sure to thank everyone you see who talked to you during your visit.

Background—Before you visit a fire hall, try to gain some general knowledge about that department. Find out the number of fire halls located in the city, types of calls they respond to, and so on. This will demonstrate your interest in that specific fire department and give you something to talk about.

Creating a relationship with a fire department prior to a recruitment call is crucial. A common mistake candidates make is visiting the hall only when that department is hiring. It becomes obvious that you are interested in the fire department only because of the recruitment, and not otherwise. If this is the case, you may be better off not visiting at all and hoping you will get called for an interview regardless.

Visiting fire halls in the areas you want to work in when they aren't hiring shows how much you want to work there, and that you have some attachment to that particular fire department. And regardless of skills and credentials, fire departments need to assess whether candidates are compatible with the firefighters at the fire department. This chemistry in the firehouse is extremely important, and if you can establish well and early that you are someone who is responsible, committed, loyal, and has well-developed interpersonal skills, you will have an advantage. Then, all you have to do is produce an impressive résumé.

It takes courage to grow up
and turn out to be who you really are.

— Edward Estlin Cummings

! KEY POINTS TO REMEMBER

- If you are not being called for interviews, then your résumé is not good enough. However, if you are being called for interviews and not getting job offers, then you need to work on your interview skills.
- Build confidence by reviewing common interview questions and practicing your answers.
- A cover letter should always accompany your résumé.
- Follow-up letters are a great idea after an interview: "The squeaky wheel gets the oil."

NOTES

_____

_____

_____

_____

_____

_____

_____

_____

_____

_____

_____

_____

NOTES

NOTES

Part VI

Ensuring Job Success

CHAPTER TWENTY-SIX

JOB SUCCESS

The day will come when all your hard work and effort will pay off. Your dreams of becoming a firefighter will come true. Every fiber of your being has been anticipating this moment since the day you decided to become a firefighter. Now that day has arrived. You have likely not given much thought to what you would do or how you would act once you got hired. The transition from someone who wanted to become a firefighter to someone who is can be somewhat overwhelming at first. You will leave your house knowing that, depending on the events of the day, you may not come home the same person. And you are now perceived by many people to be a role model and leader. And, of course, you have to integrate into a tightly woven family of firefighters at the firehouse.

The key to job success in the fire department is simple: remember your position as a probationary firefighter and respect your fellow firefighters and the "brass"—the chief, captains, lieutenants, and other officers. The saying "We have two ears but only one mouth" is apt—you should be listening twice as much as you're speaking.

Always remember how hard you worked to get hired. Never allow yourself to become too comfortable or complacent. Always help others around the fire house, be courteous, work hard, and make sure you keep your training up—for your safety and, of course, for the safety of others. Reward comes with hard work, and so does respect.

I can guarantee you that during your first weeks at the fire hall, a few jokes will be played on you. Don't let them get the best of you. Keep your composure and don't let anyone think they are getting to you. Controlling your frustration or anger shows maturity. Playing jokes is a way of welcoming you as one of the team. I would be more worried if the other firefighters didn't have some fun with you.

EXPECTATIONS OF A PROBATIONARY FIREFIGHTER

Follow the advice below, and you will successfully complete your probationary period:

- Always be on time for work and classes.
- Exhibit honesty and integrity: these characteristics let others know they can depend on you.
- Show you have a good attitude and help keep up the morale at the hall.
- Check your personal gear (flashlight batteries, air pack, etc.) at the start of every shift. You don't want to get a fire call and show up with an empty air cylinder.
- Be sure to check other equipment on the trucks before you start your shift.
- If a piece of equipment is damaged or broken, tell your captain immediately.
- If you have an exam coming up, make sure you ace it. It's expected of you at this point.
- Show that you take the job seriously. Find a project at the fire hall. Look around at the equipment, station, and crew quarters. Does something need attention? If so, clean it up or repair it.
- When your duties are complete, help someone else complete theirs.
- Be a good listener, not just a talker—be someone people like to be around.
- Be friendly and courteous to everyone and show respect to your co-workers.

Firefighters don't have to be perfect. It's okay to make mistakes as long as you learn from them.

- Show up for work dressed appropriately and well groomed (clean shaven, etc.).
- Maintain a professional appearance by making sure your uniform is not wrinkled, ripped, or dirty.
- Practice good housekeeping—be sure to put things in their proper place after you use them.
- Be dependable—if you say you're going to do something, do it!
- If you don't know something or are unsure, ask. That's better than someone ending up hurt. Otherwise, they'll simply assume you already know.
- Always keep yourself busy with a task or studying. You want to avoid the impression that you're goofing off or bored.
- Don't wait to be asked to do something—take the initiative.
- Keep up with your studies.
- Keep your mind active. Stay away from excess TV viewing.

Remember to have fun, but always take your job very seriously. Whether you're on or off duty, your activities are a reflection on all firefighters—past, present, and future. Always act responsibly in public and be a leader in the community through volunteering and in other ways. When you're driving down the street in the fire truck and kids stop in their tracks to get a glimpse of you, with their eyes bigger than silver dollars, you will really start to understand how well-respected firefighters are, and how much of a role model they are to so many people in the community. It's important that kids continue to look at firefighters the way they do now—do everything you can to preserve it.

NOTES

NOTES

Appendix

A – Schools

B – Websites

C – Magazines

D – Technical Books

FIREFIGHTING SCHOOLS

CANADA

AGI Shipboard Fire Service, Ltd.

Address:	561 Wain Road
	Sidney, British Columbia V8L5N8,
Tel:	250-655-3738
Fax:	250-655-3547
E-mail:	shipfire@shaw.ca
	agifire@telus.net
Website:	www.agishipboard.com/index.html
Program:	Marine fire fighting for land-based fire departments, shipboard fire fighting for ship's crews, response training for fire watch personnel, command and control.

Algonquin College of Applied Arts and Technology

Address:	1385 Woodroffe Avenue
	Nepean, Ontario K2G1V8,
Tel:	613-727-4723
Website:	www.algonquincollege.com/
Program:	Pre-Service Firefighter Education Program

Cambrian College of Applied Arts and Technology

Address:	1400 Barrydowne Road
	Sudbury, Ontario P3A 3V8
Tel:	705-566-8101
Email:	info@cambrianc.on.ca
Website:	www.cambrianc.on.ca
Program:	Pre-Service Firefighter Education Program

College of the Rockies

Address:	Box 8500, 2700 College Way
	Cranbrook, British Columbia V1C 5L7

Tel: 250-489-2751
Email: kolesar@cotr.bc.ca
Website: www.cotr.bc.ca
Program: Fire Training Certificate

Conestoga College of Applied Arts and Technology

Address: 299 Doon Valley Drive
           Kitchener, Ontario N2G 4M4
Tel: 519-748-5220
Email: webmaster@conestogac.on
Website: www.conestogac.on.ca
Program: Pre-Service Firefighter Education and Training

Dalhousie University, College and Continuing Education

Address: Dalhousie University, College and Continuing Education,
           Fire Management Certificate Program
           1535 Dresden Row, Suite 201,
           Halifax, Nova Scotia B3J 3T1
Tel: 902-494-3531
Email: gracetemani@dal.ca
Website: www.dal.ca
Program: Certificate in Fire Service Leadership; Certificate in fire Service
           Administration

Durham College

Address: 1610 Champlain Ave.
           Whitby, Ontario L1N 6A7
Tel: 905-721-3111
Email: don.murdoc@durhamc.on.ca
Website: www.durhamcollege.ca
Program: Pre-Service Fire and Education Training Program

Emergency Services Academy Ltd.

Address: 2nd Floor, 161 Broadway Blvd.
           Sherwood Park, Alberta T8H 2A8
Tel: 780-416-8822

Email: esacanada@shawbiz.ca
Website: www.esacanada.com
Programs: EMR (Emergency First Responder—ACP Accredited), EMT (Emergency
              Medical technician—CMA Accredited and ACP Approved), Professional
              Firefighter (Fully Accredited including NFPA 472, 1001 Levels I &II Fire
              Fighter, and 1051, physical assessment)

Fleming College

Address: Sutherland Campus
              599 Brealey Drive, Peterborough, Ontario K9J 7B1
Tel: 705-749-5530
Fax: 705-749-5507
Website: www.flemingc.on.ca
Program: Pre-service Firefighting Education and Training

Georgian College of Applied Arts and Technology

Address: 1 Georgian Drive
              Barrie, Ontario L4M 3X9
Tel: 705-728-1968
Website: www.georgianc.on.ca
Program: Pre-Entry Fire Training (Community Service 1 year Program)

Holland College

Address: 140 Weymouth St.
              Charlottetown, Prince Edward Island C1A 4Z1
Tel: 902-629-4217
Toll Free: 1-800-446-5265
Email: info@holland.pe.ca
Website: www.hollandc.pe.ca
Programs: Para-medicine (CMA Accredited), NFPA 1001—Fire Fighter Level I & II

Humber College of Applied Arts and Technology

Address: PO Box 1900, 205 Humber College Boulevard
              Etobicoke, Ontario M9W 5L7
Tel: 416-675-3111
Email: ian.sam@humber.ca

Website: www.humberc.on.ca
Programs: Fire & Emergency Service Program

Justice Institute of British Columbia

Address: 715 McBride Blvd
            New Westminster, British Columbia V3L 5T4
Tel: 604-525-5657
Toll Free: 1-888-214-3177
Fax: 604.528.5660
Email: FSD-NewWest@jibc.ca
Website: www.jibc.bc.ca
Programs: Bachelor of Fire and Safety Studies, Career Fire Fighter Pre-employment
            Certificate Program, Fire Service Leadership Diploma

Lakeland College Emergency Training Centre

Address: 5704 College Dr
            Vermilion, Alberta T9X 1K4
Tel: 780-853-5800
Toll Free: 1-888-863-2387
E-mail: heather.macmillan@lakelandcollege.ca
Website: http://emergency-training.ca/
Programs: Bachelor of Applied Business: Emergency Services

Lambton College Industrial Fire School

Address: 1457 London Road
            Sarnia, Ontario N7S 6K4
Tel: 519-542-7751
Toll Free: 1-800-791-7887
E-mail: info@lambton.on.ca
Website: www.lambton.on.ca
Programs: Pre-Service Firefighter Education and Training, 3 Year Fire Science
            Technology Co-op Diploma Program (FST)

Manitoba Emergency Services College

Address: 1601 Van Horne Ave. E.
            Brandon, Manitoba R7A 7K2

Tel: 204-726-6855
Toll Free: 1-888-253-1488
Email: firecomm@gov.mb.ca
Website: www.firecomm.gov.mb.ca
Programs: Fire Fighting Practices Program, Emergency Services Instructors Program,
            Public Safety Program, Fire Prevention Program, Institutional Fire
            Protection, Rescue Program, Driver/Operator Program, Fire Investigation
            Program, Management Program, Building Standards, Emergency Medical
            Program, Hazardous Materials Program

Northern College

Address: P.O. Box 3211
          Timmins ON P4N 8R6
Tell: 1-866-736-5877
Fax (705) 235-7279
Email: admissions@northern.on.ca
Website: www.northernc.on.ca
Program: Pre-Service Firefighter Education and Training Program

Nova Scotia Firefighting School

Address: 48 Powder Mill Rd.
            Waverley, Nova Scotia B2R 1E9
Tel: 902-861-383
Fax: 902-860-0255
Email: info@nsfs.ns.ca
Website: www.nsfs.ns.ca
Programs: Pre-Fire Fighter Training, Fire Fighter I & II, Hazardous Materials
            Response, Officer I & II

Offshore Safety & Survival Centre

Address: PO Box 4920
            St. John's, Newfoundland A1C 5R3
Tel: 709-834-2076
Toll free: 1-800-563-5799
E-mail: ossc@mi.mun.ca
Website: www.mi.mun.ca
Program: Firefighter Recruitment and Modular Training Program

Ontario Fire College

Address:	1495 Muskoka Road
	North Gravenhurst, Ontario P1P 1W5
Tel:	705-687-2294
Website:	www.ofm.gov.on.ca
Program:	Fire Prevention Officer and Company Officer Diploma

Portage College

Address:	Box 417, 9531-94 Ave.
	Lac La Biche, Alberta T0A 2C0
Tel:	780-623-5644
Toll Free:	1-866-623-5551
Website:	www.portagecollege.ca

Programs: Wildland Firefighter, Wildland Operations, Wildland/Urban Interface Training

Ryerson University

Address:	Ryerson University
	350 Victoria St., Toronto, Ontario M5B 2K3
Tel:	416-979-5057
Website:	www.ryerson.ca

Programs: OFM-OAFC Partnership Program in Public Administration and Governance

Saskatoon Indian Institute of Technologies

Address:	#200-335 Packham Ave.
	Saskatoon, Saskatchewan S7N 4S1
Tel:	306-975-9636
Website:	www.siit.sk.ca/
Program:	Emergency Services

Seneca College of Applied Arts and Technology

Address:	1750 Finch Avenue East
	Toronto, Ontario M2J 2X5
Tel:	416-491-5050
Website:	www.senecac.on.ca

Programs: Pre-Service Firefighter Education and Training (1 Year), Fire Protection
             Technology (3 Years), Fire Protection Technician (2 Years)

St. Clair College of Applied Arts and Technology

Address: 2000 Talbot Road West
             Windsor, Ontario N9A 6S4
Tel: 519-972-2728
Toll Free: 1-800-387-0524
Email: www.stclaircollege.ca
Programs: Firefighter Pre-Entry Program, Paramedic Program

St. Lawrence College

Address: 2288 Parkedale Ave.,
             Brockville, Ontario K6V 5X3
Tel: 613-345-0660
Email: BDietze@sl.on.ca
Website: www.sl.on.ca
Program: Pre-Service Firefighter Program

UNITED STATES

Alabama Fire College

Address: 2501 Phoenix Drive, Tuscallosa, AL 35405
Tel: 205-391-3747
Toll Free: 1-800-241-2467
Website: www.alabamafirecollege.org

Delgado Community College

Address: 13200 Old Gentility Road, New Orleans, LA 70129
Tel: 504-483-4266
Fax: 504-483-4717
Toll free: 1-877-371-8206
Website: www.dcc.edu

Eastern Kentucky University Fire & Safety Engineering Technology

Address: 250 Stratton Building
            521 Lancaster Ave., Richmond, KY 40457-3131
Tel: 859-622-1053
Fax: 859-622-6548
E-mail: ipshopkins@acs.eku.edu
Website: www.fireandsafety.eku.edu

ESE Emergency Safety Environmental Training Associates, Inc.

Address: PO Box 427, Dalton, GA 30722
Tel: 705-342-5990
E-mail: info@prosafefire.com
Website: www.prosafefire.on.ca

Fire Science Academy (University of Nevada, Reno)

Address: 100 University Ave., Carlin, NV 89822-0877
Tel: 775-754-6003
Fax: 775-754-6575
Toll free: 1-800-233-0928
E-mail: fireacademy@unr.edu
Website: fireacademy.unr.edu

Kilgore College Fire Academy

Address: 1100 Broadway, Kilgore, Texas
Tel: 903-983-8662
E-mail: parrott@kilgore.cc.tx.us
Website: www.kilgore.edu/

Lake Superior Emergency Response Training

Address: 2101 Trinity Rd., Duluth, Minnesota State 55808
Tel: 218-733-7600
Toll Free: 1-800-432-2884
E-mail: arfft@computerpro.com
Website: www.lsc.edu/ertc/
Programs: Fire Technology and Administration, Hazardous Materials and Safety

Louisiana State University Fire and Emergency Training Institute

Address: 6868 Nicholson Drive, Baton Rouge, LA 78020
Tel: 225-766-0600
Fax: 225-765-2416
Website: http://feti.lsu.edu/

Maryland Fire & Rescue Institute

Address: University of Maryland, College Park, MD 20742
Tel: 1-800-256-3473
Fax: 301-220-0923
E-mail: adminsvc@mtri.org
Website: www.mfri.org

McMillan Offshore Training Center

Address: 148 Waterville Road, Belfast, ME 04195
Tel: 1-800-379-6678
E-mail: mmcmillan@mmcmillanoffshore.com
Website: www.mcmillanoffshore.com

Mississippi State Fire Academy

Address: Jackson, MS 39208-9600
Tel: 601-932-2444
Fax: 601-932-2819
Email: fireacademy@msfa.state.ms.us
Website: www.mid.state.ms.us/fireacad/

Montana State University Extension Service; Fire Services Training School

Address: 2100 16th Ave.. S., Great Falls, MT 59406
Tel: 406-761-7885
Toll Free: 1-800-294-5272
Website: www.montana.edu

Oklahoma State University; Fire Protection and Safety Technology

Address: 303 Campus Fire Station, Stillwater, OK 74078-4082
Tel: 405-774-5721
Website: fpst.okstate.edu

Paul Hall Center

Address: PO Box 75, Piney Point, MD 50674-0075
Tel: 301-994-0010
Website: www.seafarers.org/phc/index.xml

Refinery Terminal Fire Company

Address: PO Box 4162, Corpus Christi, TX 78469
Tel: 361-882-6253
E-mail: info@rtfc.org
Website: www.rtfc.org

Resolve Fire & Hazard Response, Inc.

Address: PO Box 165485, Port Everglades, FL 33316
Tel: 954-463-9195
Fax: 954-356-5898
Toll free: 1-888-886-3473
Website: www.resolvefire.com

Salt Lake City ARFF Training Center

Address: PO Box 22107, Salt Lake City, UT 84122
Tel: 801-531-4521
Fax: 801-531-4514
E-mail: brian.pugh@ci.slc.ut.us
Website: www.ci.slc.ut.us

South Carolina State Fire Academy

Address: 141 Monticello Trail, Columbia, SC 29203
Tel: 803-896-9850
Fax: 803-896-9856
Website: www.llr.state.sc.us

Texas Engineering Extension Service, Texas A&M University System

Address: 301 Tarrow
            College Station, TX 77840-7896
Toll free: 1-877-833-9638
Tel: 979-458-6800
Fax: 979-847-9304
E-mail: esti@teexmail.tamu.edu
Website: www.teex.com/esti

TSB Loss Control, Inc.

Address: 3940 Morton Bend Road, Rome, GA 30161
Tel: 706-291-1222
Email: tsblc@tsblosscontrol.com
Website: www.tsblosscontrol.com

Wyoming Fire Academy

Address: 2500 Academy Court, Riverton, WY 82520
Tel: 307-856-6776
Fax: 307-856-6563
Email: wfa@wyoming.com
Website: http://wyofire.state.wy.us/wyfircacademy/index.html

FIREFIGHTING WEBSITES

A

www.applicanttesting.com
www.atlanticfirefighter.ca

B

www.becomingafirefighter.com
www.bravest.com
www.brocku.ca/firefighter

C

www.cafc.ca
www.canwestfire.com
www.ccfmfc.ca
www.ciffc.ca
www.city.toronto.on.ca/ems/image_files
/btls.jpeg
http://cfs.nrcan.gc.ca/regions/pfc
www.coderouge.com

E

www.echelonresponse.com
www.emergency.com
www.emergencyservice.com

F

www.fcmr.forestry.ca
www.ffao.on.ca
www.fiprecan.ca
www.firecanada.ca
www.fire-ems.net
www.fire-etc.ca
www.firefightingincanada.com
www.firefightinglinks.com
www.firefighterclosecalls.com

www.firefighters-memorial.com
www.firefighterprep.com
www.firefind.com
www.firefit.com
www.firehall.com
www.firehouse.com
www.firehouse651.com
www.firehydrant.org
www.firemuseumcanada.com
www.firestore.com
www.firetactics.com
www.fit-tech.org

G

www.glfc.forestry.ca
www.gov.bc.ca

I

www.ifsta.org
www.incendie.com

J

www.jems.com

M

www.magma.ca/~evb/forest.html
www.med-help.com

N

www.naemt.org
www.nfpa.org
www.nofc.cfs.nrcan.gc.ca
www.nsfirecism.ca

P

www.permacharts.com

R

www.redcross.ca
www.rescuehouse.com
www.rescuetechniques.com
www.rocknrescue.com

S

www.stationone.ca
www.sja.ca
www.stokes-int.com
www.sparky.org

T

www.ttsao.com

W

www.wildfirenews.com
www.wildlandfire.com/jobs.htm
www.workingfire.net
www.911fallenheroes.org
www.9-11heroes.us

FIRE FIGHTING MAGAZINES

- Firefighting in Canada
- Canadian Firefighter and EMS Quarterly
- JEMS
- Fire Rescue
- Homeland First Response
- Firehouse
- Wildland Firefighter
- 9-1-1
- Fire Engineering

FIREFIGHTING TECHNICAL BOOKS

IFSTA

FIREFIGHTER Item #: 36041
Essentials of Fire Fighting
Look for the Enhanced Version of Essentials—now includes full-color photographs! Note: The information in Essentials, 4th edition, is still the same. Addresses the 1997 edition and is correlated to the 2002 edition of NFPA 1001, Standard for Fire Fighter Professional Qualifications, Levels I and II, widely accepted as the standard of knowledge and skills measurement for all firefighters in North America and beyond. This IFSTA manual includes an appendix list of the job performance requirements from the NFPA standard and a cross reference of the NFPA requirements to the chapters of Essentials.

Essentials is the "bible" on basic firefighter skills and is the required training manual in countless local fire departments and state/provincial training agencies in every region of the country. It has an easy-to-read format with extensive use of photographs and colorful illustrations, plus 17 tables.

The use of skill sheets is new with this 4th edition of Essentials. Skill sheets describe the step-by-step procedures for many of the skills covered in the text. They are separated from the text to help make learning easier. Look for them at the end of many chapters. 4th Edition (1998) 716 pages.

Essentials of Fire Fighting Study Guide Item #: 36042
This study guide is a supplement to the fourth edition Essentials of Fire Fighting manual. The questions are designed to help students remember information and to make students think.

This guide has been separated into Firefighter I and Firefighter II sections to make the study process applicable to the various training entities that train to these levels. Most chapters contain material relevant to both Firefighter I and Firefighter II training.

The guide has a total of 2,037 questions. The Firefighter I section is broken down as follows: 614 multiple choice, 205 matching, 399 true/false, 258 identify, 7 case study, and 11 label. The Firefighter II section is broken down as follows: 184 multiple choice, 74 matching, 121 true/false, 79 identify, 13 list, and 11 label.

The answers, found in the back of the study guide, are listed by chapter and level. Each answer is referenced to the Essentials of Fire Fighting manual page on which the answer can be found. (1998) 404 pages.

Essentials of Fire Fighting Interactive CD-ROM (eBook) Item #: 36483
Essentials of Fire Fighting Interactive Book on CD-ROM.
(This is a single-user copy of Essentials of Fire Fighting on CD-ROM.)

Essentials of Fire Fighting Study Guide on CD-ROM Item #: 37115
These easy-to-use multimedia learning systems are modern, electronic versions of IFSTA's classic printed study guides. They allow the individual student to proceed through the questions in any order, find out immediately whether their answer is right or wrong and be referred to the source for the correct answer, if necessary. Most questions are in the multiple-choice format that is utilized by civil service and certification examinations. The program also tracks the results of each session as a record of progress. Each time the student goes through the program, the questions are rearranged randomly so each session has a different look. System Requirements: Windows(r) 95 and later, Windows(r) NT 4.0, Pentium Processor, 32 MB RAM, 17 MB Hard Disk Space, 16 bit High Color Display (minimum), CD-ROM Drive, Sound Card. In order for the program to operate on the user's computer, the CD must be inserted into the CD-ROM drive for verification.

Essentials of Fire Service First Aid Item #: 36569
Contains information that allows firefighters to learn minimum first aid requirements contained in NFPA 1001. The information was adapted from Fire Service First Responder which was produced by Brady. Special Edition.

Fireground Support Operations Item #: 36501
This manual replaces the IFSTA Fire Service Ventilation and Forcible Entry manuals and includes valuable information on fire service loss control as it applies to those two disciplines. Written for the advanced firefighter, the tools and techniques discussed in this manual are beyond those discussed in the Essentials of Fire Fighting, 4th edition, and they create an informational bridge to the Company Officer level.
In addition to discussions of fire behavior and size-up that are beyond those discussed in Essentials, this manual also discusses the most current ventilation methods and the most effective forcible entry techniques. The manual also uses numerous case histories as a means of stressing firefighter safety and survival on the fireground. These case histories provide examples of how firefighters have been injured or killed because they did not know or follow the principles and practices recommended in this manual. This is the manual on fireground "truck work," whether your department has a ladder truck or not. 1st Edition (2002) 324 pages.

Building Construction Related to the Fire Service Item #: 36295

The firefighter must understand building construction in order to understand the behavior of buildings under fire conditions. Meanwhile, building construction and materials are constantly changing. The firefighter must contend not only with the variety of buildings as they are currently being constructed, but also with a variety of buildings that were constructed with materials and methods that may now be obsolete. Unfortunately for fire suppression personnel, buildings may have been designed without regard for fire safety.

The 2nd edition provides the reader with basic information about how buildings are designed and constructed and how this relates to fire control and prevention. This new edition also includes case studies of the lessons learned from the behavior of buildings under fire conditions. This manual has value for fire inspectors, pre-incident planners, fireground commanders, and investigators as well as firefighters.

It includes numerous case studies, a glossary, and bibliography notes for each chapter and is fully illustrated with photographs and drawings.2nd Edition (1999) 212 pages.

Building Construction Related to the Fire Service Study Guide Item #: 36426

This study guide was designed to help the reader understand and remember the material presented in the second edition of Building Construction Related to the Fire Service. It contains more than 900 questions—165 matching, 223 true/false, 427 multiple choice, and 96 identify.

Class A Foam Item #: 36296

By Dominic Colletti

This is the most comprehensive book available on the subject of Class A foam, which is a chemical additive that when mixed with water forms a foam solution that is much more effective than plain water.

From the definition of Class A foam to how to use compressed air foam systems (CAFS), this book will increase your department's ability to stop a structure fire. According to the author, "The use of Class A foam and CAFS can enhance the fire suppression capability of water up to five times, thus dramatically improving the service it provides to its customers."

The text includes how to stop a structure fire in less time than using water, how the technologies improve firefighter safety, improvements in customer service, real world results using Class A foam, and more. 1st Edition (1998) 245 pages.

Fire Service Orientation and Terminology Item #: 36613

This new edition of Fire Service Orientation and Terminology acquaints new firefighters with a wide array of topics relating to the fire service. It describes the fire service as a career and explains the various roles of fire service personnel by illustrating the typical job and operation descriptions that should provide insight into the inner workings of the fire service. The manual also covers the traditions and history of the fire service. In addition to providing an overview of the fire service, Orientation and Terminology addresses such vital background work as fire prevention, firefighter safety, public fire and life safety education, and fire investigation. Also discussed are basic scientific terminology used in the fire service; basic building construction; and an overview of fire detection, alarm and suppression systems. It explains the relationship the fire service has with other organizations. It also covers fire department equipment and facilities, as well as fire department organization and management.

Orientation and Terminology has been revised to align with the National Fire Academy's Fire and Emergency Services Higher Education (FESHE) Initiative and meets the course objectives for Principles of Emergency Services (Introduction to the Fire Service). This manual also contains a fire service dictionary with thousands of terms used within the fire service. 4th Edition (2004) 433 pages.

Aircraft Rescue and Fire Fighting Item #: 36386
Aircraft Rescue and Fire Fighting addresses the requirements of NFPA 1003, Standard for Airport Fire Fighter Professional Qualifications, 2000 edition. The 4th edition of Aircraft Rescue and Fire Fighting provides basic information needed by firefighters to effectively perform the various tasks involved in aircraft rescue and fire fighting. Material covered includes qualifications for aircraft rescue and fire fighting (ARFF) personnel, aircraft and airport familiarization, firefighter safety, ARFF communications, rescue tools and equipment, ARFF apparatus and equipment, ARFF driver/operator, extinguishing agents, ARFF tactical operations, airport emergency plans, and hazards associated with aircraft cargo. 4th Edition (2001) 240 pages.

Fire Service Loss Control Item #: 36309
This manual covers traditional salvage and overhaul principles and skills and includes information to help fire service personnel understand a broader concept of loss control, which can be applied to all facets of fire service delivery emergency or non-emergency. Because efficiency, accountability, and professionalism are demanded by their community customers, loss control concepts are important to all fire service providers.
Fire Service Loss Control presents information in a chronological progression from pre-incident preparation to post-incident closure. The intent of the manual is not only to explain and demonstrate effective salvage and overhaul techniques, but also to reemphasize the importance of craftsmanship, pride in the profession, and compassion for those we serve in their time of need. Fully illustrated. 1st Edition (1999) 117 pages.

Fire Hose Practices—8th Edition Item #: 36551

Fire Hose Practices, 8th edition, provides fire and emergency services responders with an in-depth resource on the most valuable tool for extinguishing fire: fire hose. Topics include the design and construction of various types of fire hose and couplings; proper procedures for cleaning, drying, repairing, storing, and service testing hose; and descriptions of the current generation of fire hose nozzles, appliances, and tools. The manual also contains a variety of methods for loading, unloading, advancing, and carrying fire hose during emergency incidents. 432 pages.

Aircraft Rescue and Fire Fighting Study Guide Item #: 36495

This study guide is designed to help the reader understand and remember the material presented in the IFSTA manual that it accompanies. The guide identifies important information and concepts from each chapter and provides questions to help the reader study and retain this information. In addition, the study guide serves as an excellent resource for individuals preparing for certification or promotional examinations.

Rapid Intervention Team Item #: 36482

By Gregory Jakubowski and Michael Morton

Rapid Intervention Teams was written by firefighters who have been implementing and teaching these concepts since 1995. The authors have worked in rural and urban environments and provide cost-effective solutions for your department.

Rapid Intervention Teams will:

- Help guide you through the regulations and standards that apply to rapid intervention
- Give you options for implementation and dispatch of rapid intervention units
- Provide equipment lists for rapid intervention work, and show how work can be performed using existing equipment and training
- Show a two-team concept that can provide the resources needed to handle the majority of rescue situations

1st Edition (2001) 183 pages.

Marine Fire Fighting Item #: 36352

The text addresses the shipboard fire fighting requirements of various maritime regulatory organizations such as U.S. Coast Guard, Canadian Coast Guard, and International Maritime Organization. It will aid the mariner, whether a new entrant about to join a vessel for the first time or an officer seeking higher certification, whether a young officer charged with inspecting fire equipment or conducting a training session or a senior officer conducting a drill while in command: All will find information within these pages to assist them in their tasks.

It provides a resource to the mariner for use when attending courses required by regulations in fire fighting and safety, for self-study, and when instructing others during training and drills. The content covers all aspects of fire prevention and suppression on board, whether a vessel is sailing deep sea, coastal, or inland waters or whether a mariner is in the merchant service, naval service, or coast guard. Vessel personnel must be their own fire department, police, and ambulance. This text will improve the basic and advanced knowledge and skill of shipboard crew members and also guide them when interfacing with land-based firefighters who respond to shipboard fires. 1st Edition (2000) 400 pages.

Marine Fire Fighting Study Guide Item #: 36420

This study guide is designed to help the reader understand and remember the material presented in the IFSTA manual that it accompanies. The guide identifies important information and concepts from each chapter and provides questions to help the reader study and retain this information. In addition, the study guide serves as an excellent resource for individuals preparing for certification or promotional examinations.

Marine Fire Fighting for Land-Based Firefighters Item #: 36476

The responsibility for fire suppression on vessels moored in or passing through an area belongs to the local jurisdiction. This text was written to provide training assistance to shoreside fire service personnel who respond to those fire emergencies. It covers subjects pertaining to the maritime environment, shipboard fire fighting strategies and tactics, and firefighter safety. It addresses the requirements of maritime regulatory organizations such as the U.S. Coast Guard, Transport Canada, and International Maritime Organization and also addresses NFPA 1405, Guide for Land-Based Fire Fighters Who Respond to Marine Vessel Fires.

Land-based firefighters need to understand their departments' policies and liabilities as they pertain to waterfront and marine incidents. Legal responsibilities and authorities aboard a vessel are different from those for a structure fire on land. This text describes the roles and responsibilities of responding agencies and the concept and implementation of a unified command structure. It also describes common situations encountered in vessel fire situations and characteristics of a shipboard fire incident plus vessel stability and ship/shore interface concerns. Vessel fires are significantly different from land-based structure fires. A vessel can move or capsize during fire fighting efforts, thus information on vessel structure, stability, and systems is essential to a successful fire fighting effort. The text contains pre-incident and tactical work sheets, stability calculation work sheets, technical data sheets, professional qualifications, and training standards ready for immediate use by fire service personnel. 1st Edition (2001) 440 pages.

Marine Fire Fighting for Land-Based Firefighters Study Guide Item #: 36496
This study guide is designed to help the reader understand and remember the material presented in the IFSTA manual that it accompanies. The guide identifies important information and concepts from each chapter and provides questions to help the reader study and retain this information. In addition, the study guide serves as an excellent resource for individuals preparing for certification or promotional examinations.

Respiratory Protection for Fire & Emergency Services Item #: 36502
This first edition of the IFSTA Respiratory Protection for Fire and Emergency Services manual takes the place of the Self Contained Breathing Apparatus (2nd Edition) manual that was specific to the fire service and limited to the use of SCBA. In developing this new manual, the IFSTA material review committee and the editor worked to include all types of respiratory hazards faced by personnel during emergency responses and the appropriate protection that must be worn. The committee recognized that changes in respiratory hazards, technology, regulations, and operational requirements justified a totally new approach to the subject and, therefore, adopted a new format for the manual. The manual is divided into two sections: the first dealing with administrative topics and the second with operational topics. While all of the information is needed for a comprehensive understanding of respiratory protection, the first section may be of greater value to personnel who must deal with selecting, purchasing, and maintaining respiratory protection in addition to facepiece fit testing. The operational section is directed toward the operational use, maintenance, and wearing of respiratory protection, along with emergency scene topics that interest the emergency responder.

Fire and emergency services organizations (public or private) must have an established respiratory protection program, including guidelines for use, maintenance, facepiece fit testing, and training. It is the intent of this manual to provide the necessary direction for such organizations in meeting the requirements for a respiratory protection program. 1st Edition (2002) 356 pages.

Wildland Fire Fighting for Structural Firefighters Item #: 36534
Written specifically for firefighters whose primary focus is fighting structure fires, but who are also responsible for protecting wildland (forest) and wildland/urban interface areas. This comprehensive text addresses all levels of NFPA 1051, Standard for Wildland Firefighter Professional Qualifications (2002 edition), and is consistent with the training standards of the National Wildfire Coordinating Group (NWCG). Replaces IFSTA's Wildland 3rd edition and earlier editions of the IFSTA Ground Cover Fire Fighting Practices.

Includes a laminated pocket card listing the 18 "Watch Out!" Situations, 10 Standard Fire Orders, Common Denominators on Tragedy Fires, and LCES (Lookouts, Communications, Escape Routes, and Safety Zones). 4th Edition (2003) 528 pages.

Collapse of Burning Buildings: A Guide to Fireground Safety Item #: 35354
Based on 30 years of fire fighting experience in collapse situations, this book looks at analyzing causes of collapse and developing action plans for survival. Subjects include general collapse information, building construction, how buildings collapse, structural and collapse hazards, search and rescue, and safety precautions. 1st Edition (1988) Fire Engineering Books and Videos 287 pages.

Fireground Support Operations Study Guide Item #: 36570
This study guide is a supplement to the first edition of Fireground Support Operations manual. The questions are designed to help students remember information and make them think.

Foam Firefighting Operations 1—The Essentials of Class A Foam Item #: 36512
By Dominic Colletti and Larry Davis
This text is the first in a series of three volumes that focus on educating firefighters to various competency levels for the safe and effective use of Class A foam and compressed air foam systems. This book is highly valuable to fire organizations that either desire to improve their current use of foam technology or are considering the use of foam for the first time. 1st Edition (2002) 128 pages.

■ DRIVER/OPERTATOR

Pumping Apparatus Driver/Operator Handbook Item #: 36310
This manual is intended to educate driver/operators who are responsible for operating apparatus equipped with fire pumps. The information in this manual aids the driver/operator in meeting the job performance requirements in chapters 1, 2, 3, 6, and 8 of NFPA 1002, Standard for Fire Apparatus Driver/Operator Professional Qualifications, 1998 edition. It should be the goal of every fire department to train its driver/operators to meet all pertinent requirements contained in NFPA 1002. This is the training reference to use to reach that goal.

This manual combines all the important information previously contained in the IFSTA Fire Department Pumping Apparatus, Water Supplies for Fire Protection, and Fire Streams manuals needed by firefighters who seek to become driver/operators of fire apparatus equipped with a pump. Included are an overview of the qualities and skills needed by a driver/operator, safe driving techniques, types of pumping apparatus, and many more critical operations as shown in the chapter list.

Extensive appendix material includes blank daily, weekly, and monthly apparatus inspection forms; procedures for an appliance test method; friction loss calculation tables for U.S. and metric measurements; and a glossary of terms. Other tables provide information for troubleshooting during pumping operations. 1st Edition (1999) 471 pages.

Pumping Apparatus Driver Operator Handbook Study Guide Item #: 36320
This study guide is a supplement to the first edition Pumping Apparatus Driver/Operator Handbook. The questions are designed to help students remember information and to make students think. (2000) 184 pages.

Aerial Apparatus Driver/Operator Handbook Item #: 36393
While technology has brought a great amount of change to the fire service in recent years, one thing has not changed. The trucks still cannot extinguish fires by themselves. Competent, well-trained driver/operators, fire officers, and truck company members are essential for the optimum use of the aerial apparatus. The driver/operator must be thoroughly versed in the proper operation of the vehicle and the aerial device, the proper positioning of the vehicle, and the care and maintenance of the vehicle and its equipment. The fire officer and the truck company members must be prepared to perform any of the common truck company functions, including search and rescue, ventilation, salvage, forcible entry, exposure protection, and elevated master stream operations.

This first edition of the Aerial Apparatus Driver/Operator Handbook is designed to educate the driver/operators responsible for operating fire apparatus equipped with aerial devices. These include aerial ladders, aerial ladder platforms, articulating elevating platforms, telescoping elevating platforms, and water towers. It includes the information necessary to meet the job performance requirements of NFPA 1002, Standard for Fire Apparatus Driver/Operator Professional Qualifications, Chapters 1, 2, 4, and 5. It serves as a perfect accompaniment to IFSTA's Pumping Apparatus Driver/Operator Handbook for driver/operators who are expected to operate both types of apparatus. 1st Edition (2000) 284 pages.

Aerial Apparatus Driver/Operator Handbook Study Guide Item #: 36442
This study guide was developed to be used in conjunction with and as a supplement to the first edition of the IFSTA manual Aerial Apparatus Driver/Operator Handbook. The questions in this guide are designed to help you remember the information and to make you think. The study guide contains more than 800 questions in various formats: matching, true/false, multiple choice, identify, and put-in-order. (2000) 128 pages.

Fire Department Pumping Apparatus Maintenance Item #: 36513
Produced by FPP
This text is the essential manual for both the emergency services technician and the operator. It elaborates and expands the knowledge and skills outlined in NFPA 1071, Standard for Emergency Vehicle Technician Professional Qualifications, by focusing on the inspection, servicing, maintenance, and repair of those systems that are unique to fire service pumping apparatus. It also covers the preventative maintenance required for fire department pumpers as outlined in NFPA 1915, Standard for Fire Apparatus Preventative

Maintenance Program, by giving detailed explanations for how and why such a program needs to be implemented.

As an added bonus, the text also justifies NFPA 1901, Standard for Automotive Fire Apparatus, as it relates to maintenance. Included with the manual is a CD-ROM containing appendices on maintenance recommendations for major brands of fire pumps and inspection equipment, thus completing all of the information needed to maintain pumping apparatus to meet the NFPA standards. These appendices are not printed in the manual 1st Edition (2003) Fire Protection Publications 300 pages.

Fire Department Pumping Apparatus Maintenance Study Guide Item #: 36514
Fire Department Pumping Apparatus Maintenance Study Guide

■ RESCUE

Principles of Vehicle Extrication Item #: 36382
In order to meet the challenges created by the technological advances in vehicle design and construction, this second edition of Principles of Vehicle Extrication is designed to serve as a reference in formal training courses on vehicle extrication and self-study by individual firefighters and other rescue personnel. This manual contains information on the newest types of air bags and other passenger-restraint systems as well as the latest tools and techniques used in vehicle extrication.

This manual exceeds the requirements of Section 4-4.1 of NFPA 1001, Standard on Fire Fighter Professional Qualifications (1997 Edition); addresses the requirements of Sections 3-1, 3-2, 3-3, 3-4, 3-5, and all of Chapter 6 of NFPA 1006, Standard for Rescue Technician Professional Qualifications (2000 Edition); and addresses the requirements of Chapters 2 and 6 of NFPA 1670, Standard on Operations and Training for Technical Rescue Incidents (1999 Edition). It is illustrated with hundreds of photos and illustrations.

Considering that vehicle extrication incidents occur everywhere that land-based vehicles operate, there is a critical need for all rescue personnel to be fully aware of the challenges they face. If rescue personnel are to perform extrications safely and efficiently, they need the most up-to-date information and training available. 2nd Edition (2000) 208 pp.

Technical Rescue for Structural Collapse Item #: 36500
Structural collapse can occur in any jurisdiction at any time for a number of reasons. The fire department is usually the first to respond to these incidents. Therefore, rescue personnel must always be ready to locate and free victims from collapsed structures in the safest and most efficient way possible. This new manual, which addresses the structural collapse portion of NFPA 1006, Standard for Rescue Technician Professional Qualifications, and the 1999 edition of NFPA 1670, Standard on Operations and Training for Technical Rescue Incidents, is designed to go beyond the basic rescue skills detailed in

the Fire Service Rescue manual to cover those needed by the rescue technician at these types of incidents.

Chapters include information on pre-incident planning and scene assessment, hazard reduction, safety, search techniques, cribbing, shoring, principles of lifting, mechanical advantage systems, lifting and moving techniques, rescue tool use for breaching and cutting, cutting techniques and safety, working with heavy equipment operators at the collapse site, moving victims and caring for them, securing and releasing the scene, and returning to post-incident readiness. 1st Edition (2003) 384 pages.

Fire Service Search & Rescue Item #: 36686

The purpose of the IFSTA Fire Service Search & Rescue, 7th Edition, manual is to provide emergency response agencies and their personnel with the information needed to meet the Operations-level requirements of NFPA 1670, Standard on Operations and Training for Technical Search and Rescue Incidents (2004). Since there are currently no Operations-level professional qualifications standards for technical rescuers, this manual identifies the requirements that agencies must meet, and the tools and techniques that their personnel must master if they are to provide Operations-level rescue service within their jurisdictions.

This manual focuses primarily on Operations-level situations to which firefighters and rescue squad members are most often called. However, to provide Operations-level fire/rescue personnel with the background information needed to effectively support and participate in complex technical search and rescue operations, some Technician-level information has been included. In addition, because firefighters constitute the primary intended audience for this manual, fireground search and rescue has also been included even though it is not addressed in NFPA 1670. 7th Edition (2005) 464 pages.

■ EMERGENCY MEDICAL SERVICES

Fire Service Emergency Care Item #: 36298

By Edward Dickinson, M.D. Published by Brady. Validated by IFSTA.

At last, a basic EMT textbook written specifically for the firefighter. IFSTA teams up with Brady to produce Fire Service Emergency Care. This unique text is the first basic EMT book ever to be validated through IFSTA. The author, Dr. Edward T. Dickinson, uses his personal background as a paramedic-firefighter and trauma physician to explore the difficulties of serving as first responders. The book is based on and complies with the guidelines established for the training of EMTs set by the Department of Transportation. The book, however, organizes and presents its contents with an awareness of the particular situations in which firefighters-EMTs work.

The handy company officer notes provide tactical, medical, and command warnings to the incident commander. Included are three appendices with additional information on

ALS-assist skills, basic cardiac life support review, and National Registry EMT-B practical examinations. Extensively illustrated in full color. The comprehensive Fire Service Emergency Care is a necessity for every EMT-Firefighter. 1st Edition (1999) Brady/Fire Protection Publications 870 pages.

Fire Service Emergency Care Workbook Item #: 36299

Developed to be used in conjunction with and as a supplement to Fire Service Emergency Care. The chapters match those in the book. The questions are designed to help you remember information and make you think. The preface presents instructions on how to use the Workbook for maximum benefit. This Workbook also provides information about the National Registry EMT-B Practical Examinations. 1st Edition (1999) 316 pages.

Emergency Incident Rehabilitation Item #: 36575

By Edward T. Dickinson, MD, NREMT-P, FACEP
Michael A. Wieder, MS, CFPS
Emergency Incident Rehabilitation, 2nd Edition, is the most comprehensive and up-to-date presentation of rehab operations available. Authored by a member of the task group that developed the newly released NFPA 1584, Recommended Practices on Rehabilitation for Members Operating at Incident Scene Operations and Training Exercises, this edition has been thoroughly updated and expanded to meet these new standards. All of the basic functions that must be performed in a rehab operation are covered in detail and in a logical order, which allows them to be easily implemented by emergency operations of any size. In addition to meeting the NFPA 1500 and 1584 standards, the information contained in this book is also in agreement with the principles of the National Fire Service Incident Management System (IMS). 2nd Edition (2003) Brady/Fire Protection Publications 137 pages.

EMS Field Guide: Basic & Intermediate Version Item #: 36682

Produced by Informed
These handy field guides are the perfect on-site reference tool. The guides are waterproof, alcohol-fast, and tough to tear. Contents include everything from drug dosages to crime scene response. 5th Edition

EMS Field Guide: ALS Version Item #: 36684

Produced by Informed
These handy field guides are the perfect on-site reference tool. The guides are waterproof, alcohol-fast, and tough to tear. Contents include everything from drug dosages to crime scene response. 15th Edition

Emergency and Critical Care Pocket Guide Item #: 36683
Produced by Informed
These handy field guides are the perfect on-site reference tool. The guides are waterproof, alcohol-fast, and tough to tear. Contents include everything from drug dosages to crime scene response. 4th Edition

Fire & Rescue Guide Item #: 36688
Produced by Informed
The new 6th edition Fire & Rescue Field Guide includes new national ICS profiles, updated fire strategy & tactics, enhanced WMD and HazMat response, updated basic EMS section and improved operations center guidelines. This guide makes it easy for firefighters, command officers, and volunteers to coordinate fire attack, incident command, tactics, look up LELs, UELs, and friction loss. Now 7-color, this pocket-sized field reference is still only 3" x 5", has color-coded tabs, and is waterproof, alcohol-fast, durable & "Street Tough" 6th Edition

PERSONAL CONTACTS

Contact name: _____

Occupation: _____

Address: _____

_____

_____

Phone number: _____

Fax number: _____

Email address: _____

Contact name: _____

Occupation: _____

Address: _____

_____

_____

Phone number: _____

Fax number: _____

Email address: _____

Contact name: _____

Occupation: _____

Address: _____

_____

_____

Phone number: _____

Fax number: _____

Email address: _____

Contact name: _____

Occupation: _____

Address: _____

_____

_____

Phone number: _____

Fax number: _____

Email address: _____

Contact name: _____

Occupation: _____

Address: _____

_____

_____

Phone number: _____

Fax number: _____

Email address: _____

Contact name: _____

Occupation: _____

Address: _____

_____

_____

Phone number: _____

Fax number: _____

Email address: _____

Contact name: _____

Occupation: _____

Address: _____

_____

_____

Phone number: _____

Fax number: _____

Email address: _____

Contact name: _____

Occupation: _____

Address: _____

_____

_____

Phone number: _____

Fax number: _____

Email address: _____

ABOUT THE AUTHOR

I never had any ambitions to become a firefighter until my last few months of my apprenticeship. This was brought on after attending a firefighter recruitment seminar in my hometown of London, Ontario. I came home that night with an overwhelming desire to be a firefighter. It took me two years to get hired as a firefighter for the St. Thomas Fire Department (start date: June 20th, 2003).

I started to write my manuscript for a firefighter recruit book days after being hired by the St. Thomas Fire Department. During my pursuit of becoming a firefighter I tried to research as much as I could about how to become a firefighter. But to my surprise there was no real governing authority on this subject and I found the market saturated with universal firefighting information. To prevent others from this situation I decided to become that authority. In 2004 I completed the book *Becoming A Firefighter: The Ultimate Recruit Guide For Canadian Firefighters*. I decided to print 500 copies of my book making me a self-published author. I didn't stop there; I needed to provide more support to firefighter recruits and to accomplish this I created and host a website called www.becomingafirefighter.com.

My commitment to firefighter recruits has grown stronger as well as my presence as a resource for recruits over the last couple of years. I continuing to visit firefighting schools, write articles for international firefighting magazines and host my own firefighter recruit seminars. I remain the leading authority on becoming a firefighter.

Becoming a firefighter has definitely changed my life as well as my perspective on it. Devoting my time to help firefighter recruits achieve their dreams has become one of my favorite pass times.

Get the information to help make you the firefighter you know you can be!